Optimisation of ZnO Thin Films

Saurabh Nagar · Subhananda Chakrabarti

Optimisation of ZnO Thin Films

Implants, Properties, and Device Fabrication

Springer

Saurabh Nagar
Department of Electrical Engineering
Indian Institute of Technology Bombay
Mumbai, Maharashtra
India

Subhananda Chakrabarti
Department of Electrical Engineering
Indian Institute of Technology Bombay
Mumbai, Maharashtra
India

ISBN 978-981-10-0808-5   ISBN 978-981-10-0809-2 (eBook)
DOI 10.1007/978-981-10-0809-2

Library of Congress Control Number: 2017938554

Printed on acid-free paper

This Springer imprint is published by Springer Nature
The registered company is Springer Nature Singapore Pte Ltd.
The registered company address is: 152 Beach Road, #21-01/04 Gateway East, Singapore 189721, Singapore

Preface

Extensive research is currently carried out on ZnO as a potential material for the fabrication of optoelectronic devices such as laser diodes (LDs) and light-emitting diodes (LEDs) in the ultraviolet (UV) region because of its wide bandgap (3.437 eV at 2 K) and a large excitonic binding energy of 60 meV at room temperature. However, the bottleneck in ZnO-based devices is the creation of reliable and reproducible p-type films, because ZnO is an intrinsically n-type material due to defects such as oxygen vacancies and zinc interstitials. The low solubility of dopants and self-compensating process on doping further aggravate the problem. This monograph describes the different implantation mechanisms which can be used to achieve strong, reliable and stable p-type ZnO thin films which would result in revelations in the optoelectronics field in the UV region. This monograph will be very useful for new doctorant students who want to work on doping and implantation of ZnO thin films and subsequently fabricating optoelectronic devices. Chapter 1 of the monograph emphasises the importance of ZnO in the field of optoelectronics for ultraviolet (UV) region and also discusses the material, electronic and optical properties of ZnO. Chapter 2 discusses the optimisation of the pulsed laser-deposited (PLD) ZnO thin films in order to make successful p-type films and in order to achieve the high optical output required for higher efficiency devices. It also discusses a hydrogen implantation study on the optimised films to confirm whether the implantation leads to improvement in the optimised results. Chapter 3 gives details about the various implantation studies performed on ZnO thin films in order to achieve successful and reliable p-type ZnO films. Conventional ion implantation (CII) was used to implant lithium and phosphorus ions, while plasma immersion ion implantation (PIII) was used to implant phosphorus and nitrogen ions. Once the p-type ZnO was achieved, the fabrication and characterisation of the different ZnO-based optoelectronic devices were carried out which have been discussed in Chap. 4. The successful fabrication of heterojunction devices such as p-ZnO/n-Si diode and p-ZnO/p-Si detectors is demonstrated. Fabrication of p-ZnO/n-ZnO homojunction LED is also demonstrated. The last chapter summarises the various results of the experiments presented and provides an insight into potential further studies in this field.

We would like to thank Prof. David C. Look of Wright State University for sharing his knowledge, which has helped us in better understanding of the subject. We would also like to thank Dr. Suhas Jejurikar, Dr. Nilanjan Halder and Dr. Bhavesh Sinha for their fruitful discussions when they were in IIT Bombay. We are thankful to Dr. S.K. Gupta, Mr. Arindam Basu and Mr. B.V. Subhramanyam for helping me out in performing the LEAF experiments. We are also thankful to the SPM facility at IIT Bombay for performing the AFM measurements. We are thankful to all our present and past group members Dr. Arjun Mandal, Dr. Saumya Sengupta, Dr. Sourav Adhikary, Sandeep Kala, M. Kulasekaran, Aijaz Ahmed, Saikalash Shetty, Hemant Ghadi, Goma KC and Akshay Balgarkashi. We would like to thank all the laboratory members and facility staff of IITBNF. We are grateful to our funding agencies namely Department of Science and Technology, Govt of India, and MCIT through IIBNF.

Keywords—II–VI semiconductors, ZnO, LEDs, Detectors, Pulsed laser deposition, Sputtering, Low-Energy Accelerator Facility, Plasma immersion ion implantation, X-ray diffraction, Atomic force microscopy, Scanning electron microscopy, Van der Pauw Hall, Photoluminescence, Thermally stimulated current, Electroluminescence, Spectral response.

Mumbai, India

Saurabh Nagar
Subhananda Chakrabarti

Contents

About the Authors

Saurabh Nagar received his Bachelor of Technology in electronics and communication engineering from the Bengal College of Engineering and Technology, West Bengal University of Technology, India, and completed his Ph.D. from the Department of Electrical Engineering, Indian Institute of Technology Bombay, Mumbai, India, in 2013. He worked as a Research Associate at the Indian Institute of Technology Bombay for 9 months followed by a Postdoctoral Fellow at Institut des Nanotechnologie de Lyon (INL), Ecole Centrale de Lyon, for 18 months. Presently, he is a scientific employee at the Institute Functional Oxides for Energy Efficient Information Technologies at Helmholtz-Zentrum Berlin. His research interest includes ZnO-based optoelectronics devices and oxide-based materials for memristor applications.

Subhananda Chakrabarti received his M.Sc. and Ph.D. degrees from the Department of Electronic Science, University of Calcutta, Kolkata, India, in 1993 and 2000, respectively. He was a Lecturer in the Department of Physics, St. Xavier's College, Kolkata. He has been a Senior Research Fellow with the University of Michigan, Ann Arbor, from 2001 to 2005; a Senior Researcher with Dublin City University, Dublin City, Ireland, from 2005 to 2006; and a Senior Researcher (RA2) with the University of Glasgow, Glasgow, UK, from 2006 to 2007. He joined as an Assistant Professor in the Department of Electrical Engineering, IIT Bombay, Mumbai, India, in 2007. Presently, he is a Professor in the same department. He is a Fellow of the Institution of Electrical and Telecommunication Engineers (IETE), India, and also a Member of the IEEE, MRS (USA), SPIE (USA), etc. He is the 2016 medal recipient of the Materials Research Society of India and was also awarded the 2016 NASI-Reliance Industries Platinum Jubilee Award for application-oriented innovations in physical sciences. He serves as an Editor of the IEEE Journal of Electron Device Society. He has authored more than 250 papers in international journals and conferences. He has also co-authored a couple of book chapters on intersubband quantum dot detectors.

Dr. S. Chakrabarti serves as reviewer for a number of international journals of repute such as Applied Physics Letters, Nature Scientific Reports, IEEE Photonics Technology Letters, IEEE Journal of Quantum Electronics, Journal of Alloys and Compound and Material Research Bulletin. His research interests lie in compound (III–V and II–VI) semiconductor-based optoelectronic materials and devices.

Abbreviations

ABE/A°X	Acceptor-bound-exciton
AFM	Atomic force microscope
Al	Aluminium
ALD	Atomic layer deposition
Au	Gold
CII	Conventional ion implantation
CVD	Chemical vapour deposition
DAP	Donor-to-acceptor pair
DBE/D°X	Donor-bound exciton
EL	Electroluminescence
FA	Free electron-to-acceptor
FWHM	Full width at half maximum
FX	Free exciton
GaN	Gallium nitride
IPA	Isopropyl alcohol
ITO	Indium tin oxide
k/k_B	Boltzmann constant
KrF	Krypton Fluoride
LD	Laser diode
LEAF	Low-Energy Accelerator Facility
LED	Light-Emitting Diode
LO	Longitudinal optical
MBE	Molecular beam epitaxy
NBE	Near-band-edge emission
Ni	Nickel
NiO	Nickel oxide
PIII	Plasma immersion ion implantation
PL	Photoluminescence
PLD	Pulsed laser deposition
RF	Radio frequency

rms	Root mean square
RTEL	Room temperature electroluminescence
SEM	Scanning electron microscope
SIMS	Secondary ion mass spectroscopy
TCE	Trichloroethylene
Ti	Titanium
TSC	Thermally stimulated current
UV	Ultraviolet
V_O	Oxygen vacancy
XRD	X-ray diffraction
Zn_i	Zinc interstitials
ZnMgO	Zinc magnesium oxide
ZnO	Zinc oxide

List of Figures

List of Tables

Chapter 1
Introduction

Abstract Extensive research is currently carried out on ZnO as a potential material for the fabrication of optoelectronic devices such as photodiodes, laser diodes (LDs) and light-emitting diodes (LEDs) in the ultraviolet region because of its wide bandgap (3.437 eV at 2 K) and a large excitonic binding energy of 60 meV at room temperature. However, because of intrinsic defects such as oxygen vacancies and zinc interstitials, ZnO is intrinsically deposited as an n-type material. Being a II–VI semiconductor compound, group I and group V elements are considered suitable p-type dopants for ZnO. In this monograph, authors concentrate mostly on the different methods undertaken to achieve strong, reliable and stable p-type ZnO films. Ion implantation, being a suitable technique to achieve selective and localised doping, was used to dope ZnO with different group I and group V elements. Lithium and phosphorus ions were implanted using the conventional ion implantation technique, while plasma immersion ion implantation was used to implant phosphorus and nitrogen ions in the ZnO thin films. Such studies would bring interesting revelations in the optoelectronics field for the ultraviolet range and help to produce devices for the commercial market at a much cheaper cost.

Keywords II–VI semiconductor · ZnO · LED's · Detectors

In recent years, researchers across the globe have shown great interest in zinc oxide (ZnO) as a semiconductor material, primarily due to its interesting prospects in the field of optoelectronics applications owing to its wide and direct bandgap ($\sim$3.437 eV at 8 K) [1]. In fact, ZnO has become a strong competitor of gallium nitride (GaN), another wide bandgap semiconductor ($\sim$3.4 eV at 300 K), as they share most common properties. However, due to its large exciton-binding energy of 60 meV (GaN: 21–25 meV), the devices fabricated from ZnO have a much greater probability of surviving at higher temperature. Moreover, ZnO is highly resistant to radiation damage required for space applications. Finally, in terms of suitability for wet etching, the availability of high-quality ZnO bulk crystals and the potentially lower cost for ZnO-based devices, ZnO ranks higher than GaN as well as other

S. Nagar and S. Chakrabarti, *Optimisation of ZnO Thin Films*,
DOI 10.1007/978-981-10-0809-2_1

transparent conductors like indium tin oxide (ITO). Therefore, ZnO is capable of replacing its competitors in the near future [2, 3].

ZnO growth is possible with most known processes, including chemical vapour deposition (CVD), molecular beam epitaxy (MBE), RF sputtering and pulsed laser deposition (PLD), among others. Due to its large exciton-binding energy, it can also be used for near-band edge emission (NBE) at room temperature and higher temperatures. ZnO has also been a suitable candidate for space applications due to its high resistance to radiation damage. Recently, ZnO has also found applications in transparent thin film transistors [1].

1.1 Material Properties

Most of the II–VI binary compound semiconductors crystallise as either zinc blende or hexagonal wurtzite structure where each anion is surrounded by four cations or vice versa. The crystal structure of ZnO is wurtzite (B4), zinc blende (B3) and rock salt (B1), as shown in Fig. 1.1. Of these, wurtzite structure is the thermodynamically stable phase in ambient conditions. The zinc blende structure is stabilised only by growth on cubic structures, while rock salt structure can be obtained at high pressures [1].

The wurtzite structure has a hexagonal unit cell with lattice parameters, a and c. Figure 1.2 shows the wurtzite structure of ZnO [4]. The structure is composed of two hexagonal-close-packed sub-lattices, each of which consists of one type of atom displaced with respect to each along the c-axis [1].

As mentioned earlier, ZnO has many important properties that make it a deserving candidate for primacy in the semiconductor industry. Some of its properties are wide bandgap, large exciton-binding energy and high resistance to radiation damage, among others. ZnO has a melting point of 1975 °C and is insoluble in water and alcohol, but can be dissolved in dilute acids. Some of the other properties of ZnO are given in Table 1.1 [5].

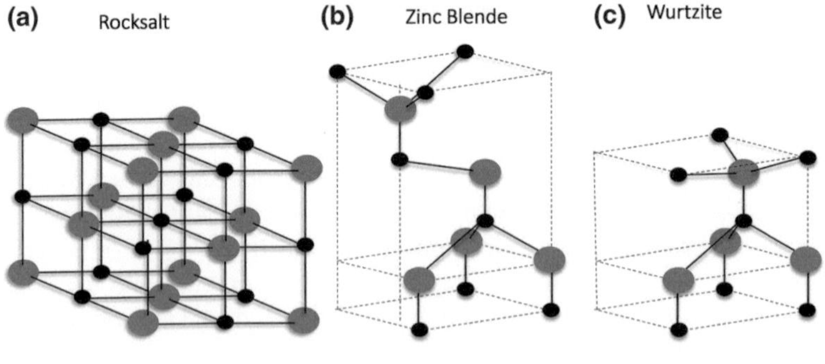

Fig. 1.1 Representation of ZnO crystal structures: **a** rocksalt, **b** cubic zinc blende and **c** wurtzite. The gray and black spheres denote Zn and O atoms respectively

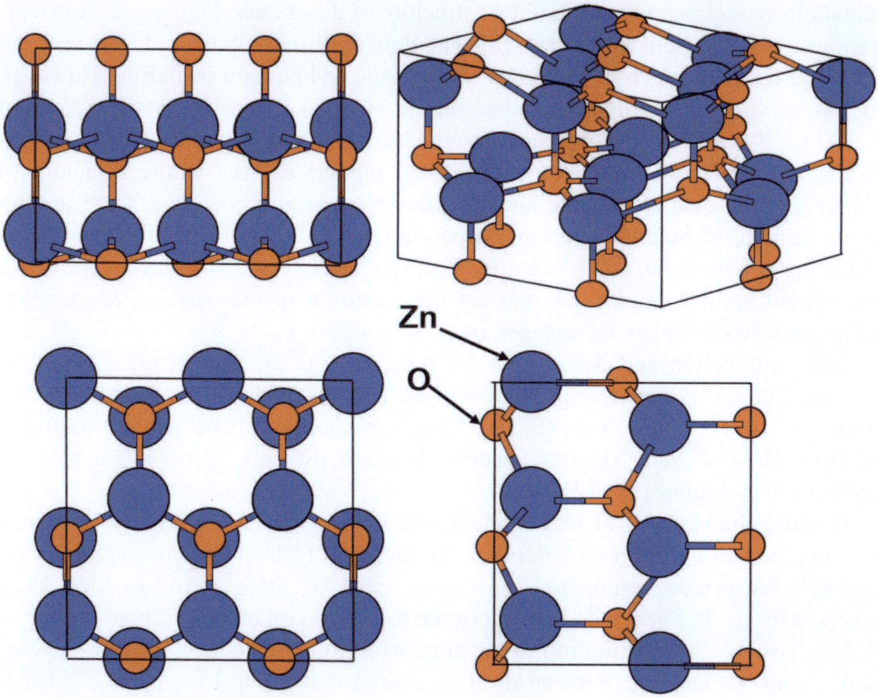

Fig. 1.2 Crystalline forms of ZnO in the wurtzite structure

Table 1.1 Properties of ZnO

Properties	Specifications	Comments
Structure	Wurtzite	Same as hexagonal
Bandgap	3.3 eV	Wide bandgap semiconductor
Exciton-binding energy	60 meV	Higher than GaN (21–25 meV)
Piezoelectricity	1.2 cm^{-2}	Highest among all semiconductors
Hardness	~5 Ω	For single crystal
Toxicity	Very low	Same as other compound semiconductors

1.2 Electronics Properties

ZnO is intrinsically an n-type semiconductor material due to the existence of native defects such as zinc interstitials (Zn_i) and oxygen vacancies (V_O), as well as the presence of impurities in the form of hydrogen, which act as donor states in ZnO [6, 7]. A vacancy is an unoccupied lattice site, which results in an unsatisfied bond within lattice. A vacancy is created when species leave lattice sites and move to the surface. Interstitial defects are present if species do not occupy the lattice site but remain in crystal and disturb the ideal structure of the crystal. Figure 1.3 shows the various defects present in a crystal [8]. The bulk electron mobility is in the range of 200–400 cm^2/Vs, which is limited by the acoustic and phonon scattering. However, in the case of thin films, the electron mobility is typically in the range of 1–100 cm^2/Vs due to the interface roughness and grain boundary scattering [3]. A major obstacle in fabricating ZnO-based bipolar devices is the difficulty of achieving high-quality, reliable and reproducible p-type ZnO. Thus, there are not much data on the hole transport in p-type ZnO due to the problems associated with obtaining p-type ZnO. This lack of data explains why the commercial usage of ZnO-based bipolar devices has not yet been realised in the research community, although it has so many advantages over other materials.

The introduction and activation of p-type dopants are technically quite challenging. The self-compensating process of the dopants tends to transform the p-type films back to n-type over time [9, 10]. Moreover, the dopants have a low solubility in ZnO. Thus, forming shallow acceptor levels is difficult, which aggravates the problem of p-doping [11, 12].

Because ZnO is a II–VI semiconductor material, group I and group V elements are supposed to be the p-type dopants for ZnO and form shallow acceptor levels, required for p-type conduction. However, most of these elements form deep acceptor levels and hence do not contribute to p-type conduction. p-type doping in ZnO is possible by substitution of Zn sites with group I elements or oxygen sites with group V elements. Park et al. [13] showed that group I elements are better p-type dopants than group V elements, in terms of forming shallow acceptor levels. However, they tend to occupy interstitial sites rather than substitutional sites and thus act like donors [14]. Nevertheless, researchers around the world have reported p-type ZnO films by doping group I elements such as lithium (Li), sodium (Na) and potassium (K) [15–22]. Due to the large size of the group V elements, the theory

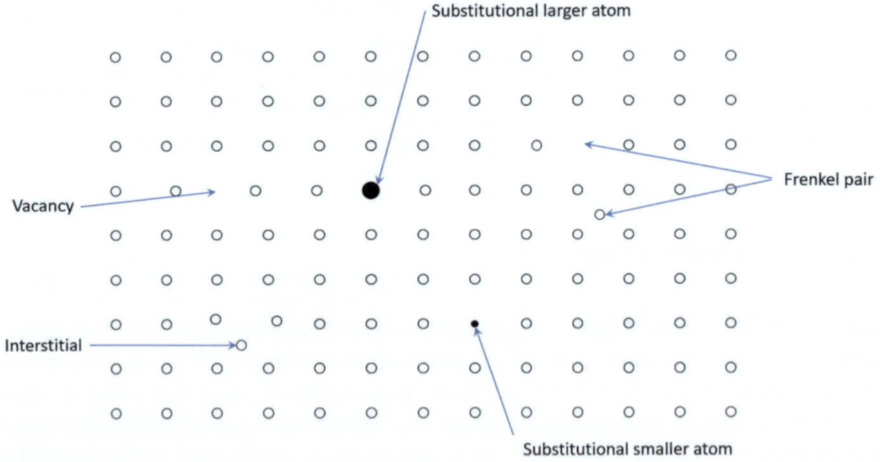

Fig. 1.3 Defects in crystals

suggests that it is difficult to achieve shallow acceptor levels [13]. Among the group V elements, nitrogen (N) is best suited to form p-type ZnO thin films because of its similar atomic size compared to oxygen. Substituting oxygen atoms by nitrogen doping, the acceptor activation energy is predicted to be around 100 meV, which is among the smallest. However, Li et al. [23] revealed that stability using nitrogen doping is an issue due to compensation of acceptor states by hydrogen. Nonetheless, successful p-type ZnO has been reported by various researchers across the world [24–34]. p-type ZnO has also been reported by other group V elements such as phosphorus (P) [35–42], arsenic (As) [9, 43–50] and antimony (Sb) [46, 51–59]. As these elements have very large atomic size compared to oxygen, substitution by these atoms tends to form deep acceptor levels at around 200 meV. Limpijumnong et al. [46] have reported the formation of a defect complex with the dopant atom and the Zn site, e.g. As_{Zn}–$2V_{Zn}$, which forms shallow acceptor levels and contributes to p-type doping. Apart from these dopants, p-type ZnO has also been reported by co-doping methods. In this method, a donor–acceptor (group I and group III, or group III and group V) elements are co-doped together [60–67].

Most of these reported results of p-type ZnO films have been achieved using non-localised mechanisms, which is not desirable for device fabrication. Ion implantation is a suitable technique for attaining selective and localised doping. Although researchers have successfully demonstrated p-type ZnO films using ion implantation technique, the existing reports are minimal [68–71]. In this report, implantation technique has been applied in order to achieve high-quality, reliable and reproducible p-type ZnO films.

1.3 Optical Properties

Intrinsic and extrinsic effects strongly affect the optical properties of a semiconductor. Intrinsic transitions take place between electrons in conduction band and holes in valence band. These also include excitonic transition due to Coulomb interaction. Extrinsic transitions relate to defects and dopants, which create discrete electronic states in the bandgap. The electronic states of bound excitons depend upon the type of semiconductor. Thus, excitons can be bound to neutral or charged donors or acceptors. Hence, there are mainly three types of transitions, i.e. shallow neutral-donor-bound exciton (DBE/D°X), shallow neutral-acceptor-bound-exciton (ABE/A°X) and defect-related transitions like free-to-bound (electron-acceptor) or bound-to-bound (donor–acceptor) [72].

The conduction band of wurtzite ZnO is mainly constructed from s-like state, whereas the valence band is created from p-like states, which are split into three sub-bands (A, B and C) due to the influence of crystal fields and spin-orbit coupling [1]. The free exciton (FX) transitions are related to transitions from conduction band to a valence band, or vice versa, and Teke et al. [72] observed this peak around 3.3771 eV. In any direct bandgap semiconductor, DBE/D°X and ABE/A°X are the most important extrinsic transitions. The neutral shallow DBE dominates in

ZnO because of the presence of donors, due to unintended impurities such as Zn_i and V_O. ABE peaks are observed when acceptors are included in the sample. The D °X peaks for the A excitons are positioned at 3.3598, 3.3605, 3.3618, 3.3650 and 3.3664 eV. Higher D°X peaks are accredited to excited-state transitions. The A°X peaks are found to be present at 3.3564, 3.3530 and 3.3481 eV [72]. Other than these important transitions, transitions like electron-acceptor (FA) peak are observed in the range 3.3–3.33 eV, whereas donor–acceptor peaks are found at 3.21–3.23 eV. Yellow/green luminescence is also observed in samples having very high defect density. These peaks correspond to deep-level transitions of the sample due to its defects [72].

1.4 Light Emitters

The wide bandgap and large exciton-binding energy of ZnO make it a very promising candidate for fabrication of light emitters, which requires that the junction devices provide high electron and hole concentration in the active region. While electron concentration is easier to achieve, the requirements of achieving high hole concentration make the device fabrication difficult. Hence, different materials having high hole concentration are used as other materials to fabricate heterojunction devices [73–75]. However, some researchers also report fabrication of homojunction light emitters [48, 76, 77].

1.4.1 Heterojunction Light-Emitting Diodes

To fabricate a heterojunction device, two issues need to be considered: (1) band offsets at the heterojunction (Fig. 1.4) and (2) defects at the heterojunction interface. The band offsets are given by [3]:

$$\Delta E_c = E_c(\text{n-ZnO}) - E_c(\text{p-emitter}) \tag{1.1}$$

and

$$\Delta E_v = E_v(\text{p-emitter}) - E_v(\text{n-ZnO}) \tag{1.2}$$

In the case of ZnO junction light emitters, the valence band offset is the limiting factor because of the presence of large energy barriers to hole injection. For this to take place, the hole concentration should be larger than the electron concentration and ΔE_c should be larger than ΔE_v. The defects at the interface occur due to

Fig. 1.4 Band diagram of a ZnO junction diode illustrating carrier recombination processes with electrical injection

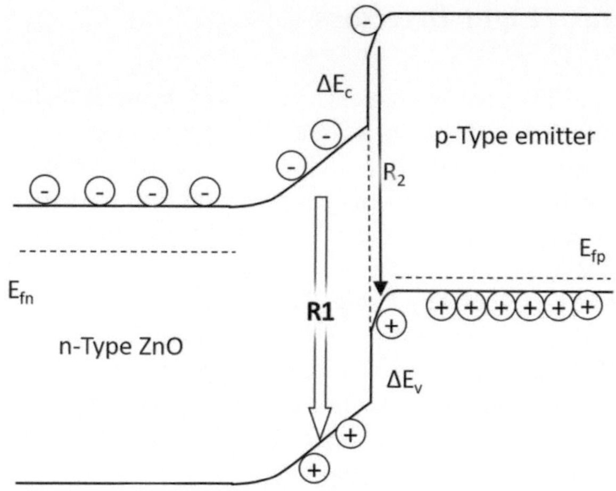

differences in crystal structure and chemical bonding. To reduce these differences, a high crystal quality at the active region is required.

Among the p-emitters, GaN is the most widely used material, because the crystal structure and electronic properties of GaN and ZnO are very similar. Moreover, high-quality p-type GaN can easily be made. Alivov et al. [78] successfully fabricated p-GaN/n-ZnO heterojunction LED where they observed an electroluminescence (EL) peak around 430 nm, which matched the cathodoluminescence spectrum of GaN:Mg, suggesting that the active region is in the GaN region. To shift the active region to ZnO, they further modified the structure by fabricating a n-ZnO/p-$Al_{0.12}Ga_{0.88}N$ heterojunction [79]. Other researchers have also fabricated heterojunction devices based on GaN [80–82]. Researchers have also successfully demonstrated the use of heterojunction devices by using other p-type materials such as Si, SiC and other oxides such as NiO and $SrCr_2O_2$ [73–75, 83–86].

1.4.2 Homojunction Light-Emitting Diodes

Several reports have been published regarding the achievement of p-type ZnO films, but realisation of room-temperature electroluminescence (RTEL) based on ZnO p–n homojunction diode is quite limited. Tsukazaki et al. [76] successfully demonstrated a RTEL of a ZnO homojunction LED structure. The EL demonstrated a peak at 440 nm corresponding to donor-to-acceptor peak. However, the emission of green band near 2.2 eV is larger than the blue band, which Tsukazaki et al. [87] eliminated by optimising the growth parameters. Liu et al. [48, 88] and Lim et al. [77] also demonstrated RTEL dominated by deep-level emission. Lim et al. further improved the device structure to achieve UV emission.

1.5 Light Detectors

Spectroscopic instruments, used for chemical and biological analysis as well as solar blind imaging, require UV light sources and detectors for its efficient functioning [3]. Although light detection is possible using semiconductors having bandgap less than the targeted photons, for efficient UV detection, it is preferable to detect only the desired UV photons rather than other shorter wavelengths. Hence, UV detectors are generally fabricated from wide and direct bandgap semiconductors corresponding to the UV spectral region. Researchers use ZnO to fabricate UV detectors because of its useful electronic and optical properties in the UV region. Photoconductors are the simplest types of detectors in which the conductivity of the material changes when photo-generated current is added [89]. Due to its simplicity, ZnO-based photoconductors are the most widely studied device [90–95]. Schottky or ohmic contacts are deposited on ZnO thin films to complete the device. Another type of detector is the junction detector or a photodiode, which is typically a p–n junction (heterojunction or homojunction) device. This device has the advantage of having lower dark current as opposed to photoconductors. The device works in the reverse bias mode: under illumination, the photo-generated carriers are quickly swept out of the depletion region, due to the electric field resulting in an increased reverse bias current [89]. Difficulty in p-doping of ZnO has restricted the fabrication of homojunction photodiodes, although Ryu et al. [96] did demonstrate one of the homojunction photodiodes. However, demonstration of several heterojunction photodiodes has been reported using Si, NiO and ZnTe [3, 97–101].

Although p-type ZnO thin films were reported before this monograph was started, the reliability of the obtained p-type ZnO thin films had not been investigated. Hence, this work was undertaken in order to produce reliable and reproducible p-type ZnO over a long period of time, which would further support the fabrication of stable ZnO-based devices required for commercial production. For successful commercialisation of these devices, optical efficiency should be as high as possible. Thus, methods to improve the optical efficiency of the ZnO thin films were also studied herein.

Chapter 2
Optimisation of PLD Parameters

Abstract Progress in the fabrication of ZnO-based optoelectronic devices lies in producing reproducible, reliable and stable p-type ZnO films, because of the intrinsic n-type nature of the deposited ZnO films. Hence, for successful conversion of its carriers from n-type to p-type, it is desirable that the deposited ZnO film has as low an electron concentration as possible. Moreover, for the fabrication of optoelectronic devices, the films should have very high optical quality. The growth of highly oriented films with the least strain will offer an added advantage in fabricating these devices. Hence, the deposition parameters of temperature and pressure using pulsed laser deposition (PLD) were optimised keeping these things in mind. A substrate temperature of 650 °C and oxygen pressure of 40 mTorr were found to be optimised growth parameters as it had the lowest carrier concentration of 1.01×10^{17} cm^{-3} a reasonably high Hall mobility of 16.1 cm^2 V^{-1} s^{-1} and also had the highest optical quality. Once the PLD parameters were optimised, hydrogen implantation was carried out to see whether it further enhanced the electrical and optical properties of the thin film. While the Van der Pauw Hall measurements did not reveal any significant changes in the electrical characteristics of the thin films, the optical quality of the implanted films was found to increase by two orders of magnitude when compared to the as-deposited sample. Such an enhancement in the optical luminescence of the ZnO thin films may be helpful in fabricating highly efficient ZnO-based devices.

Keywords PLD · X-ray diffraction · Atomic force microscopy · Scanning electron microscopy · Van der Pauw Hall · Photoluminescence · Hydrogen implantation · Thermally stimulated current

Part of the chapter has been reprinted from [S. Nagar et al. "Increased photoluminescence of hydrogen-implanted ZnO thin films deposited using pulsed laser deposition technique", Journal of Luminescence, Vol. 15, Pg 307–311, 2014 (with permission from Elsevier)]

S. Nagar and S. Chakrabarti, *Optimisation of ZnO Thin Films*,
DOI 10.1007/978-981-10-0809-2_2

2.1 Introduction

Progress in the fabrication of ZnO-based optoelectronic devices lies in producing reproducible, reliable and stable p-type ZnO films, because of the intrinsic n-type nature of the deposited ZnO films. Hence, for successful conversion of its carriers from n-type to p-type, it is desirable that the deposited ZnO film has as low an electron concentration as possible. Moreover, for the fabrication of optoelectronic devices, the films should have very high optical quality. The growth of highly oriented films with the least strain will offer an added advantage in fabricating these devices.

Deposition of ZnO thin films was carried out using PLD over ⟨001⟩ sapphire substrates. A very low lattice mismatch between ZnO and sapphire substrates explains the use of the substrate. In this study, we optimise (i) the substrate temperature and (ii) the oxygen pressure, keeping in mind the above-mentioned requirements for fabrication of devices.

The deposited films were characterised in order to observe the structural, electrical and optical properties of the films with an intention to obtain films with the least strain, defects and carrier concentration and the highest optical efficiency. The structural properties were characterised using XRD and SEM, while the electrical and optical properties were measured with the Van der Pauw Hall technique and PL, respectively.

2.2 Temperature Optimisation

2.2.1 Experimental Details

ZnO thin films were deposited over ⟨001⟩ sapphire substrates using PLD technique. The 99.999% pure ZnO target was ablated using a KrF excimer laser (248 nm with a 20-ns pulse) for the deposition. The energy density and the target-to-substrate distance were maintained at 1.94 J cm^{-2} and 5 cm, respectively. To attain uniform deposition and avoid pit formation, the target was rotated at a frequency of 10 Hz, the same as the laser frequency. Prior to deposition, a high vacuum of 3×10^{-5} mTorr was kept in the chamber, and the substrate was degreased by treatment with trichloroethylene (TCE), acetone and isopropyl alcohol (IPA) for 2 min each. The deposition was performed in a high vacuum chamber under a background oxygen pressure of 0.1 mTorr for 25 min by varying the temperature from 400 to 800 °C.

2.2.2 Results and Discussion

2.2.2.1 Structural Properties

The X-ray diffraction pattern for the film deposited at 650 °C is as shown in Fig. 2.1. Peaks at 34° and 72° could be clearly seen which corresponds to the ⟨002⟩ and ⟨004⟩ ZnO peaks which suggest that ZnO films grow with a highly c-axis

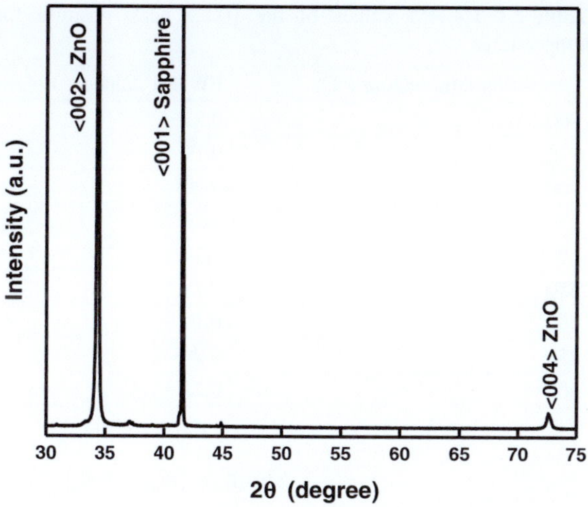

Fig. 2.1 X-Ray Diffraction spectrum of sample deposited at 650 °C, showing the ⟨002⟩ and ⟨004⟩ ZnO peaks

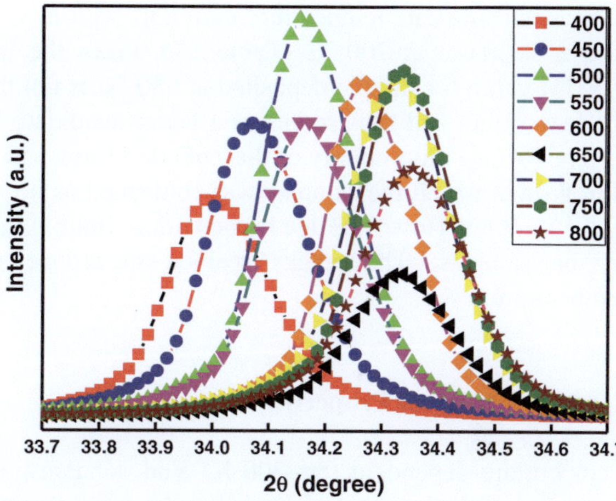

Fig. 2.2 ⟨002⟩ peak positions for different samples

orientation. Absence of other ZnO peaks indicates high crystalline quality of the deposited films which is not the case for films deposited by other techniques like spray pyrolysis or Atomic Layer Deposition (ALD) [102, 103]. It is to be noted that similar spectra was observed for samples deposited at other temperatures. However, a closer examination of the ⟨002⟩ peak positions for different samples (Fig. 2.2) displays that the peaks were at different 2Θ angles for different samples. This was due to the lattice strain, calculated by the following equation, caused by the mismatch between the lattice constants of ZnO and sapphire.

Table 2.1 FWHM values of the ⟨002⟩ peak position for samples deposited at different temperatures

Deposition temperature (°C)	FWHM values (°) (correct up to 2 decimal places)
800	0.22
750	0.19
700	0.20
650	0.20
600	0.18
550	0.19
500	0.19
450	0.20
400	0.22

$$\text{Strain} = (d_{\text{subs}} - d_{\text{film}})/d_{\text{subs}} \tag{2.1}$$

where d_{subs} and d_{film} are the *d*-spacing of the substrate and the film, respectively. Further, the FWHM values of the ZnO peaks was found to decrease with an increase in substrate temperature until 650 °Cm then increase slightly, followed by further decrease at 700 °C (Table 2.1). From the table, it can be seen that the FWHM value for sample deposited at 650 °C is not the lowest, but this sample has the least strain and thus serves as a better candidate for device applications.

The surface morphology of the ZnO thin films was studied using SEM. Epitaxial film growth for all the samples was confirmed as no grain formation was observed in the samples. However, due to the lattice strain in the films, cracks were observed in the samples. The cracks seemed to reduce with increase in deposition temperature.

2.2.2.2 Electrical Properties

Temperature-dependent (80–300 K) Van der Pauw Hall measurements were performed to study the effect of deposition temperature on the electrical properties of the films (Figs. 2.3 and 2.4). The electron concentration was found to decrease from 3.1×10^{18} cm^{-3} at 800 °C to 1.49×10^{18} cm^{-3} at 650 °C followed by further increase to 4.23×10^{18} cm^{-3} at 400 °C at 400 °C at 300 K (Table 2.2). The corresponding Hall mobility was calculated to be 15.7, 24 and 8.69 cm^2 V^{-1} s^{-1}, respectively. Although no major changes in the electron concentration values was found for any of the samples with increase in temperature, a slight increase in the Hall mobility values was identified. As there was no grain formation in the films due to the epitaxial growth of the films, the grain boundary scattering, which is generally affected by temperature, was negligible, thus explaining the increase in mobility values. The measurements revealed that all the films have n-type characteristics.

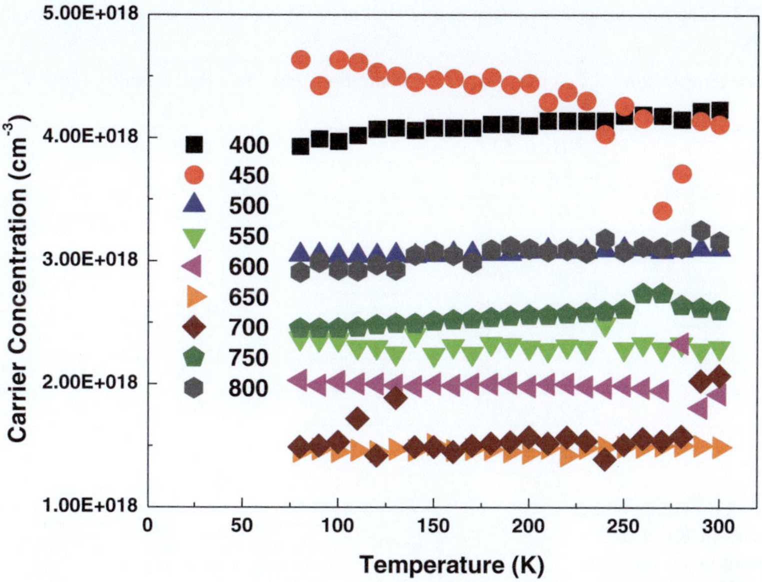

Fig. 2.3 Variation of electron carrier concentration with temperature from 80 K to 300 K

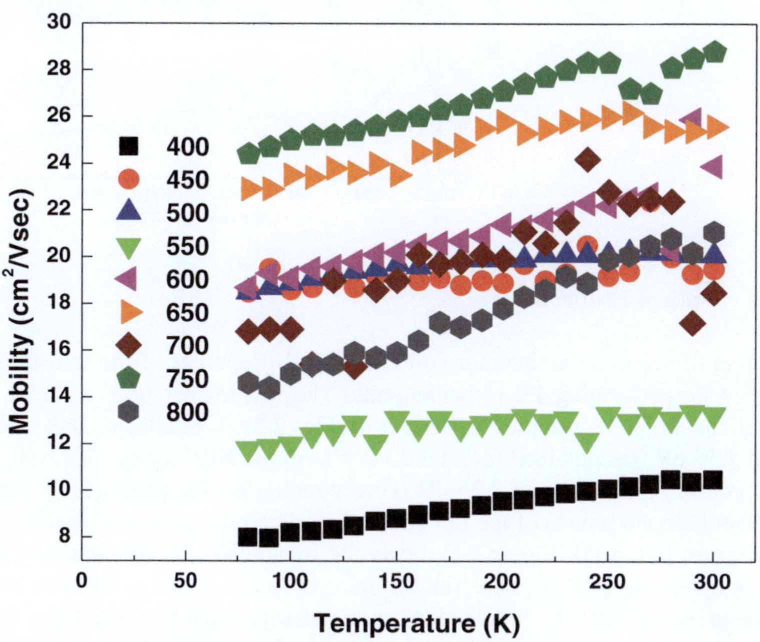

Fig. 2.4 Variation of electron Hall mobility with temperature from 80 to 300 K

Table 2.2 Electron carrier concentration and Hall mobility values for the obtained samples taken at 300 K

Substrate temperature (°C)	Electron carrier concentration (cm^{-3})	Hall mobility ($cm^2\ V^{-1}\ s^{-1}$)
	300 K	300 K
800	3.16×10^{18}	21.1
750	2.6×10^{18}	28.8
700	2.07×10^{18}	18.5
650	1.49×10^{18}	25.6
600	1.92×10^{18}	23.9
550	2.3×10^{18}	13.3
500	3.1×10^{18}	20.1
450	4.11×10^{18}	19.5
400	4.23×10^{18}	10.5

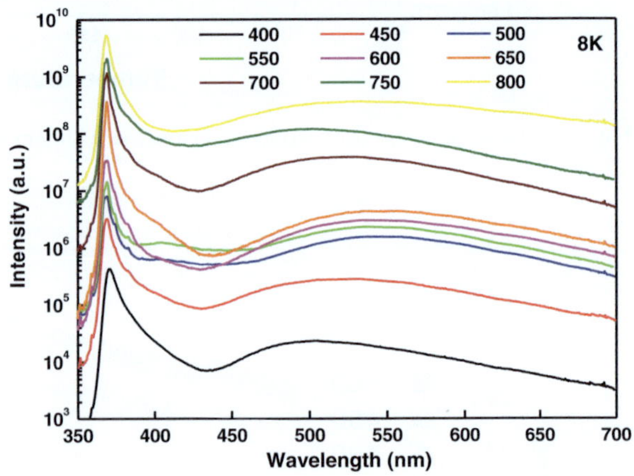

Fig. 2.5 Low temperature PL spectra (8 K) of the samples showing NBE and deep-level emissions

2.2.2.3 Optical Properties

The effect of substrate temperature on the optical properties of the thin films was measured by performing PL measurements. The low-temperature (8 K) PL spectrum of the different samples is shown in Fig. 2.5. A dominant D°X emission around 3.36 eV was obtained for all the samples in the NBE region [72, 104]. Other peaks around 3.37, 3.33 and 3.25 eV corresponding to FX, peak due to extended defects and second replica of the LO phonon, respectively, were also observed from the spectrum [72, 105]. Figure 2.6 shows the integrated PL peak intensity of different samples for the D°X peak. The PL intensity is first found to decrease at lower temperatures until 550 °C, followed by an increase to reach a maximum at 650 °C. Further increase in the deposition temperature leads to a decrease in the PL intensity. Along with the NBE peaks, deep-level defect peaks centred around 2.4 eV, leading to yellow and green emissions are also detected. These defect peaks

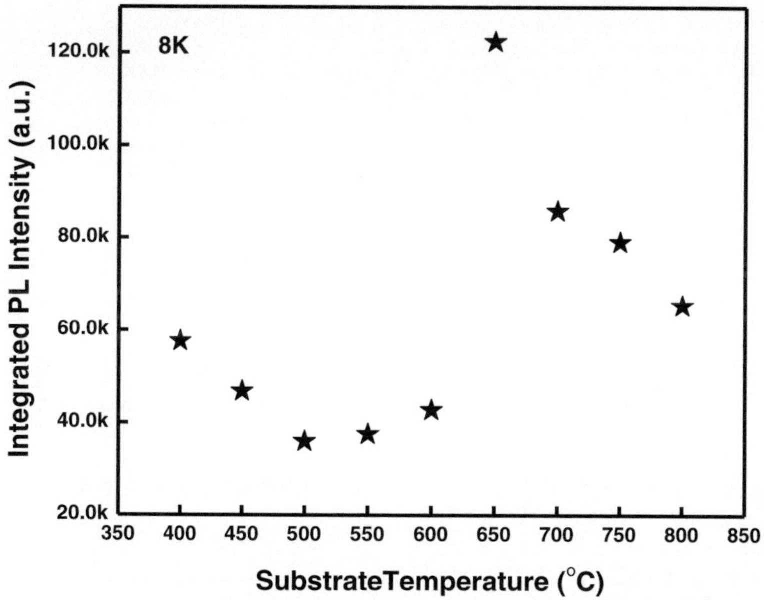

Fig. 2.6 Integrated PL peak intensity for the D°X peak of all samples taken at 8 K

arise due to intrinsic defects such as Zn_i (yellow emission) and V_O (green emission). A closer look at these peaks suggest that the green emission occurs at higher temperatures (>650 °C), while the yellow emission is seen at low and intermediate substrate temperatures. However, the intensity of these defect peaks is much lower when compared to the NBE peaks because of the growth of stoichiometric thin films from PLD which is not the case for films grown by other methods [106, 107].

Figure 2.7 shows the temperature-dependent spectrum for a sample deposited at 650 °C. A dominant D°X and a FX shoulder peak are observed at lower temperatures, as mentioned above. At higher temperatures, the FX peak starts to dominate the D°X peak. The peak intensities of FX and D°X with temperature are shown in Fig. 2.8. While the D°X peak diminishes with temperature and vanishes above 150 K, the FX peak intensity first increase and then starts to decrease. The thermal dissociation of bound excitons into free excitons at higher temperatures leads to such an observation [108]. Decrease of the FX peak with temperature >150 K is due to the increase in the number of defects with temperature. The evolution of the PL peak energies for the FX and D°X with temperature is shown in the inset of Fig. 2.8. The band gap shrinkage of ZnO is calculated by fitting the temperature dependence of FX peak with temperature using Varshni's equation:

$$E_g(T) = E_g(0) - \alpha T^2/(T + \beta) \tag{2.2}$$

where $E_g(0)$ is the band gap of ZnO at $T = 0$ K, and α and β are the fitting parameters calculated to be 7×10^{-4} eV K^{-1} and 515 K, respectively. The band

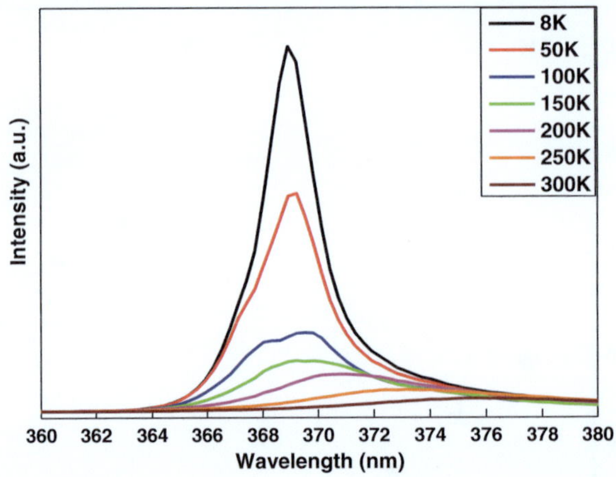

Fig. 2.7 Temperature-dependent Photoluminescence spectrum for the sample deposited at 650 °C

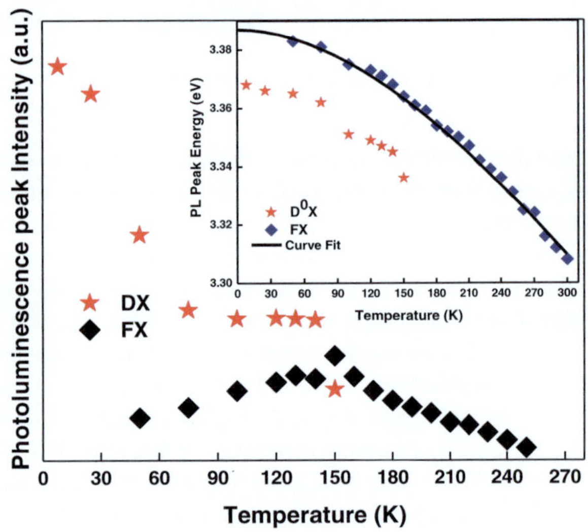

Fig. 2.8 PL peak intensity variation with temperature for the FX and D°X peaks deposited at 650 °C. *Inset* temperature dependence using Varshni's law

gap of the ZnO film is found to be 3.387 eV, which is in agreement with previously reported values [72, 109].

2.2.3 Conclusion

To summarise, PLD technique was used to deposit ZnO thin films over ⟨001⟩ sapphire substrates by varying the substrate temperature from 400 to 800 °C. XRD and SEM results confirmed deposition of highly c-axis-oriented ⟨002⟩ epitaxial ZnO films for all the samples. Temperature-dependent Van der Pauw Hall measurements were carried out in order to achieve the optimised sample having the lowest electron carrier concentration and high mobility values for obtaining better p-type films in future. The sample deposited at 650 °C was found to give the lowest carrier concentration of 1.49×10^{18} cm^{-3} and a high Hall mobility value of 24 cm^2 V^{-1} s^{-1}. Temperature-dependent PL measurements were carried out to study the optical properties of the thin films and also to attain a film with highest optical quality. Sample deposited at 650 °C was found to be the optimum sample as it had the lowest strain, carrier concentration, a reasonably high Hall mobility value and the best optical quality. As the substrate temperature increases, the ZnO atoms get sufficient energy to diffuse across the substrate, thus resulting in much uniform and high-quality film. However, increasing the substrate temperature too high results in out-diffusion of the ZnO atoms from the surface of the substrate which again degrades the quality of the film. It can be noticed from the various structural, electrical and optical properties that at 650 °C, the ZnO atoms get sufficient energy to diffuse across the substrate uniformly without out diffusing form the surface, thus resulting in the optimised parameters.

2.3 Pressure Optimisation

Once the temperature of deposition was optimised, it was required that the oxygen pressure is also optimised so that the carrier concentration becomes minimal along with increased optical luminescence observed from PL spectra. For this purpose, ZnO thin films were deposited over ⟨001⟩ sapphire substrates using the same energy density, target-to-substrate distance and deposition time as mentioned in Sect. 2.2.1. The optimised substrate temperature of 650 °C was used for deposition. The oxygen pressure in this case was varied from 0.1 to 60 mTorr.

2.3.1 Results and Discussion

2.3.1.1 Structural Properties

XRD results verify the deposition of (002) ZnO thin films with peaks around 2θ angle of 34° and 72° corresponding to the ⟨002⟩ and ⟨004⟩ ZnO peaks (Fig. 2.9). The (002) peak shows a shift in position for the various samples, which may be due to the strain developed in the films. The strain in the films has been calculated using Eq. 2.1 and has been plotted for different samples (Fig. 2.10). It is observed that the strain first increases with oxygen pressure up to 10 mTorr, followed by a decrease to attain a minimum value at 40 mTorr. Upon further increasing the oxygen pressure, the strain is found to increase.

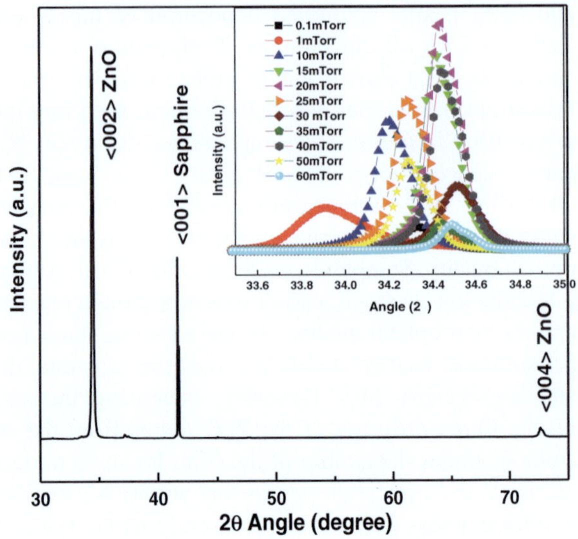

Fig. 2.9 XRD image of the sample deposited at 40 mTorr showing the ⟨002⟩ and ⟨004⟩ ZnO peaks. *Inset* shows the ⟨002⟩ peak positions of all the samples

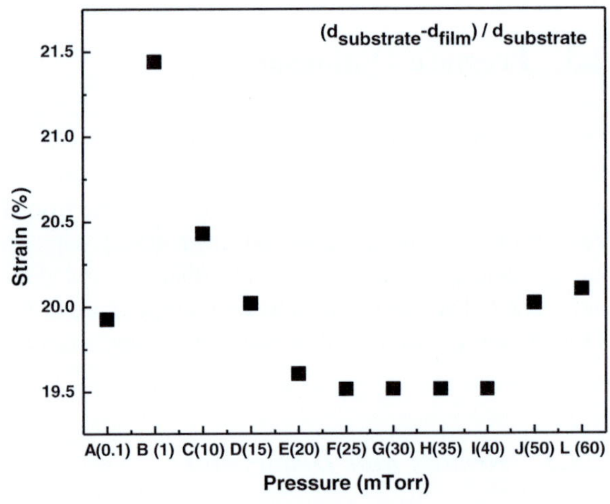

Fig. 2.10 Variation of lattice strain of ZnO thin films with oxygen pressure

SEM images of the ZnO surface are shown in Fig. 2.11. As the oxygen pressure is increased, grain formation starts to take place. Moreover, the grains are not continuous at lower oxygen pressure (up to 15 mTorr). When the pressure is further increased, the grains grow denser continuously, corresponding to the high quality of the films. However, increasing the pressure beyond 40 mTorr again leads to less dense films and discontinuous grains. A possible reason might be the collisions of the atoms at high pressure during deposition.

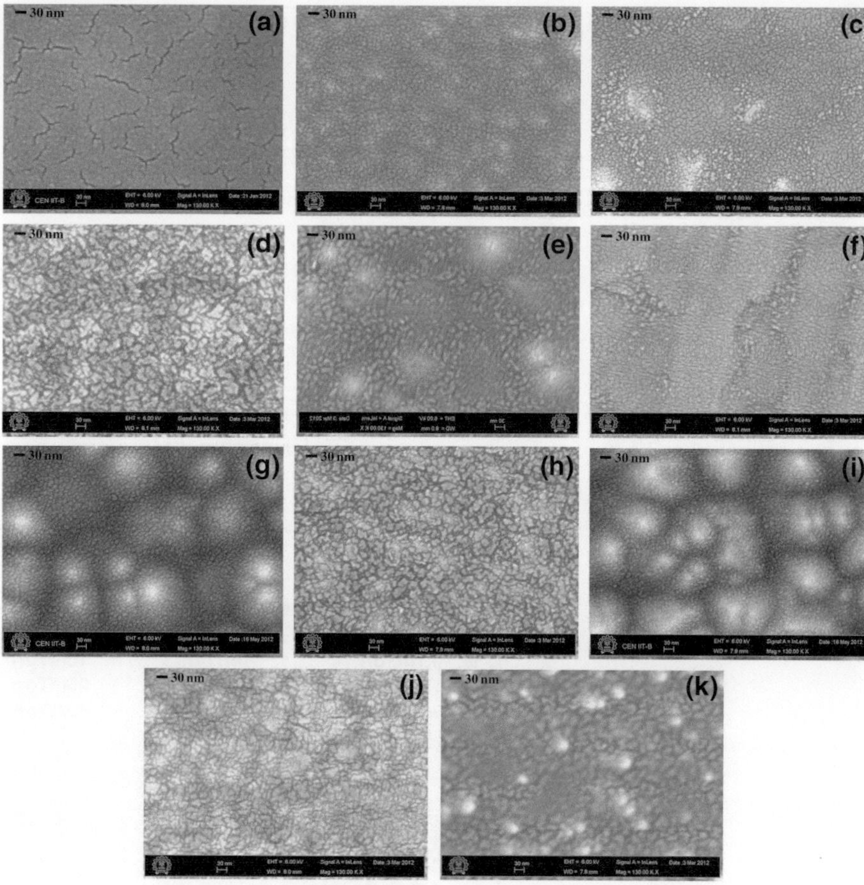

Fig. 2.11 Scanning electron microscopy (*SEM*) images of different samples: **a** 0.1, **b** 1, **c** 10, **d** 15, **e** 20, **f** 25, **g** 30, **h** 35, **i** 40, **j** 50 and **k** 60 mTorr, respectively

2.3.1.2 Electrical Properties

Temperature-dependent (80–300 K) Van der Pauw Hall measurements were performed to study the electrical properties of the ZnO thin films (Figs. 2.12 and 2.13). Table 2.3 shows the carrier concentration and Hall mobility values at 80 and 300 K. The electron carrier concentration decreases with increase in oxygen pressure, because the amount of V_O reduces as the pressure increases. However, at higher oxygen pressures (>40 mTorr), the value again increases by an order of magnitude. This may be because less energy is transferred to the substrate due to increased collision among atoms. The Hall mobility values also decrease with pressure until 25 mTorr, after which it starts increasing to gain a sufficiently high value of 16.1 cm^2/V s. The high mobility value at 0.1 mTorr may be due to the lack of grain formation in the sample, leading to negligible grain boundary scattering. In

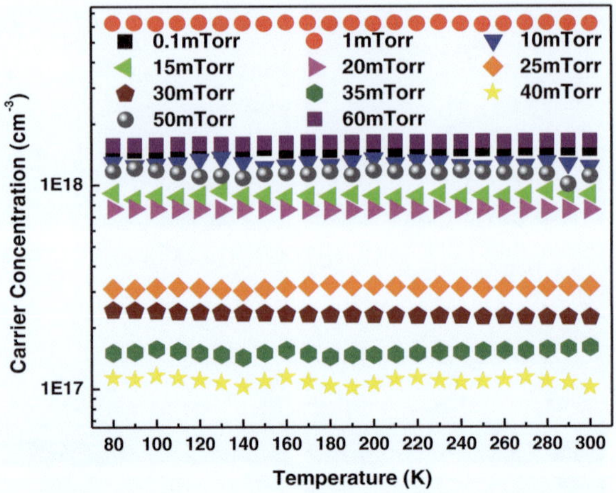

Fig. 2.12 Variation of carrier concentration with temperature for various samples

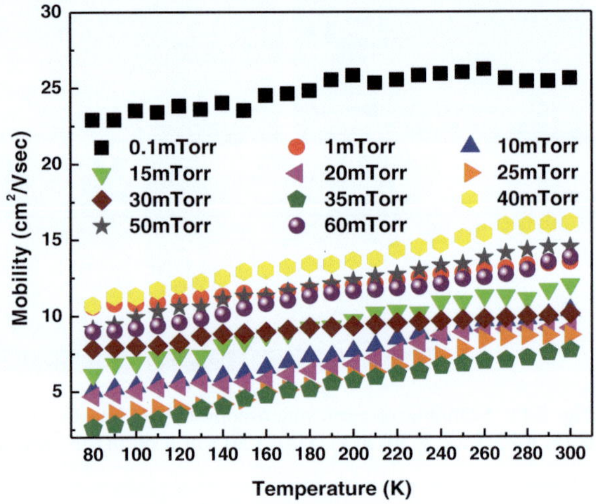

Fig. 2.13 Variation of Hall mobility with temperature for different samples

this case, too, the variation of carrier concentration with temperature is not observed. However, the Hall mobility tends to show an increase in value with temperature.

2.3.1.3 Optical Properties

Figure 2.14 shows the PL spectra of different samples taken at 8 K of the NBE region. A dominant D°X peak is visible for all the samples around 3.36 eV, along with FX peak at 3.37 eV. The other peaks at 3.33 and 3.25 eV are due to extended defects and LO-phonon replica [72, 104, 105]. Deep-level defect peaks around

Table 2.3 Carrier concentration and Hall mobility values for the obtained samples at 80 and 300 K

Pressure (mTorr)	Carrier concentration (cm^{-3})		Hall mobility (cm^2/V s)	
	80 K	300 K	80 K	300 K
0.1	1.45×10^{18}	1.49×10^{18}	24	25.6
1	6.2×10^{18}	6.14×10^{18}	10.6	13.5
10	1.3×10^{18}	1.23×10^{18}	5.05	10.3
15	9.12×10^{17}	8.92×10^{17}	6.14	12
20	7.72×10^{17}	7.5×10^{17}	4.67	9.15
25	3.1×10^{17}	3.16×10^{17}	3.36	8.71
30	2.43×10^{17}	2.2×10^{17}	7.83	10.2
35	1.5×10^{17}	1.58×10^{17}	2.61	7.7
40	1.13×10^{17}	1.01×10^{17}	10.7	16.1
50	1.17×10^{18}	1.1×10^{18}	9.15	14.5
60	1.57×10^{18}	1.64×10^{18}	8.97	13.8

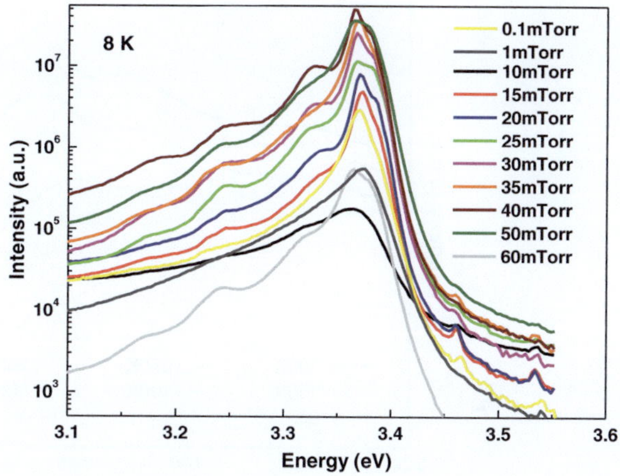

Fig. 2.14 PL spectra of the samples taken at 8 K showing NBE transition

2.5 eV are also observed in the samples, although the intensity of the peak is quite low when compared to the NBE peak. This suggests that the ZnO films thus grown are of high quality. The integrated PL peak intensity for D°X peak of different samples is shown in Fig. 2.15. The intensity first decreases with pressure until 40 mTorr, followed by an increase, to attain a highest value at 40 mTorr oxygen pressure. The increased oxygen available during deposition causes fewer defects related to V_O, causing the film to be of higher quality. However, increasing the oxygen pressure further leads to increased collision among atoms, thus reducing the peak intensity.

The temperature-dependent spectra for a sample grown at a pressure of 40 mTorr are shown in Fig. 2.16. A dominant D°X peak is observed at lower

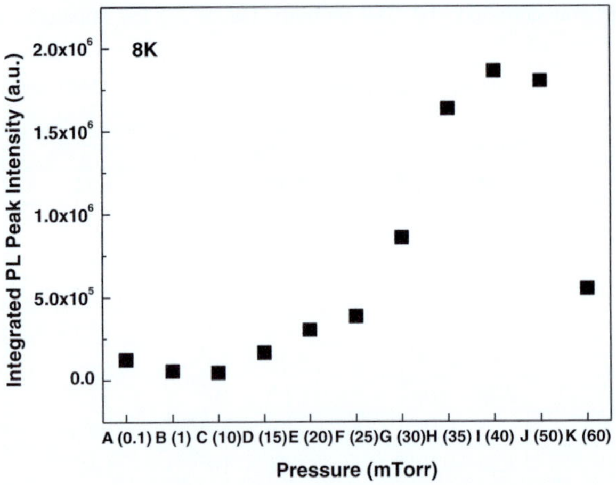

Fig. 2.15 Integrated PL peak intensity at 8 K for the D°X peak of all samples

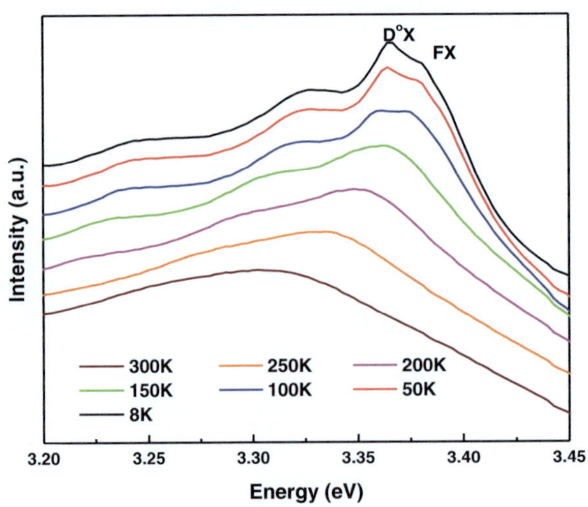

Fig. 2.16 Temperature-dependent PL spectra for the sample grown at 40 mTorr

temperatures with an FX shoulder. Enhanced thermal dissociation of bound exciton to FX leads to domination of FX peak over D°X peak with temperature [108]. Figure 2.17 shows the FX and D°X PL peak intensities with temperature. A rapid decrease in the D°X peak is observed, which vanishes entirely beyond 150 K. The FX peak, on the other hand, first increases up to 50 K, after which it starts to decrease. Inset of Fig. 2.17 shows the evolution of the PL peak energy with temperature for the D°X and FX peaks, which can be fitted using Eq. 2.2.

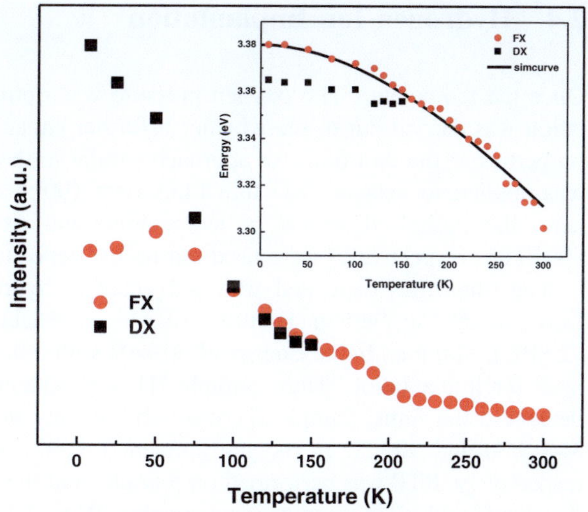

Fig. 2.17 Variation of PL peak intensities with temperature for the FX and D°X peaks for sample deposited at 40 mTorr. *Inset* temperature dependence using Varshni's law

2.3.2 Conclusion

ZnO films were deposited over ⟨001⟩ sapphire substrates at an optimised temperature of 650 °C in different oxygen pressure. Highly c-axis-oriented ⟨002⟩ epitaxial films were obtained for all the samples. The sample deposited at a pressure of 40 mTorr was found to have the lowest strain and very high grain density, as observed from SEM. Temperature-dependent Van der Pauw Hall measurements within a range of 80–300 K were obtained to achieve an optimal sample with low carrier concentration and high mobility values. The lowest concentration of 1.01×10^{17} cm^{-3} at 300 K was obtained with a corresponding Hall mobility of 16.1 cm^2 V^{-1} s^{-1} for a sample deposited at an oxygen pressure of 40 mTorr. Temperature-dependent PL spectra were also examined to study the optical behaviour of the films and to obtain a sample with high optical quality. The oxygen pressure of 40 mTorr yielded an optimal sample with the least strain and carrier concentration, along with a high Hall mobility value and the best optical quality. It is a well-known fact that an increase in the pressure of a chamber reduces the mean free path of the particle. Hence, at low pressures, the mean free path is more which corresponds to the atoms in the plume travelling more distance, and thus, the plume does not converge at the substrate leading to non-uniformity and degraded film quality. Similarly, too high a pressure causes to plume to converge well before the substrate, and hence, most of the atoms are unable to get deposited on the film again leading to poor film quality and non-uniformity. From the results obtained through the various characterisation techniques, it is believed that at an oxygen pressure of 40 mTorr, the plume converges at the substrate leading to high-quality films.

2.4 Hydrogen Ion Implantation

Once the temperature and oxygen pressure was optimised, hydrogen ion implantation was carried out to see whether it further enhanced the electrical and optical properties of the thin film. An approach similar to the one described in Sect. 2.2.1 was applied to deposit ZnO thin films over ⟨001⟩ sapphire substrate. However, here, the optimised values of temperature and oxygen pressure (650 °C and 40 mTorr, respectively) were used during the deposition time to obtain Sample A.

The films thus deposited were subjected to hydrogen ion implantation using Low-Energy Accelerator Facility (LEAF) at Bhabha Atomic Research Centre (BARC), Mumbai, India. Energy of 50 keV with a dose of 5×10^{12} ions/cm^2 was used for implantation. Thus, Sample H1 was obtained. As implantation leads to defects in the films, Sample H1 was subsequently rapid thermal annealed at 750, 800, 850 and 900 °C in oxygen ambient for 30 s to produce Samples H2–H5, respectively. RTP was performed on Sample A at the same temperatures and using the same annealing parameters (samples B to E) to compare them with the implanted samples.

2.4.1 Results and Discussion

2.4.1.1 Structural Properties

XRD was performed on the obtained samples to study the effect of hydrogen implantation and subsequent annealing on the structural properties of the thin films (Fig. 2.18). The results confirm the deposition of highly c-axis-oriented ⟨002⟩ ZnO films for all the samples. No other peaks have been found in the XRD peaks, which suggest that ZnO preferentially grow in the ⟨001⟩ direction due to its lowest surface free energy. A closer look at the ⟨002⟩ peak for the different samples shows that the Samples H1–H5 are at a higher 2θ angle as compared to Sample A. The strain developed due to the hydrogen implantation and subsequent annealing causes this effect (Fig. 2.19). The strain was found to increase for Sample H1, which is due to the implantation-related defects. The reduction of defects on annealing leads to lower strain in these samples, although the strain is still higher when compared to Sample A.

2.4.1.2 Electrical Properties

Room temperature Van der Pauw Hall measurements were taken in order to visualise the effect of implantation on the electrical properties on the thin films (Table 2.4). A decrease in the carrier concentration and Hall mobility is observed for Sample H1 as compared to Sample A. The heating caused by hydrogen

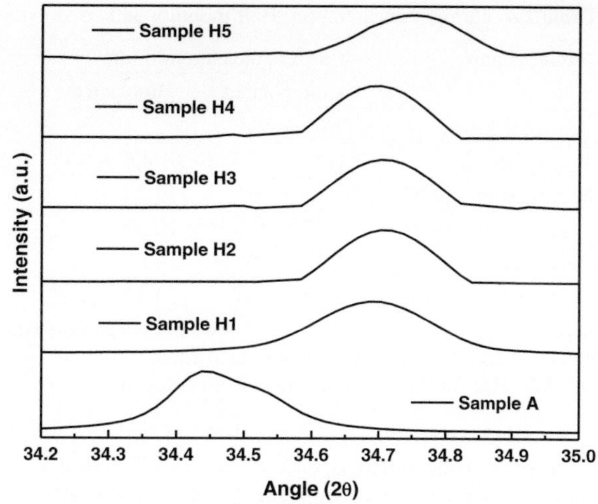

Fig. 2.18 XRD spectra of the different samples showing the ⟨002⟩ ZnO peak. The peak shifts for the implanted samples are clearly visible

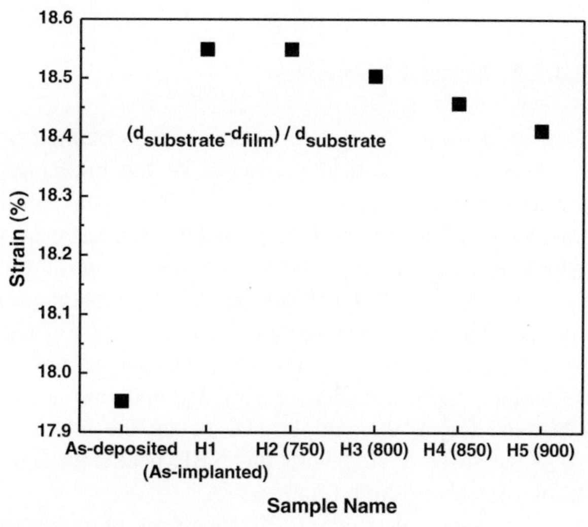

Fig. 2.19 Strain on the ZnO films due to implantation

implantation may have reduced the intrinsic defects in the sample, leading to lower concentration. However, the increased number of atoms in the films leads to higher scattering, thus reducing the Hall mobility. Upon annealing the samples, at higher temperatures, the carrier concentration and Hall mobility start to increase.

Hall measurements were obtained for Samples B to E to check whether hydrogen ions have any effect on the electrical properties of the thin films. Interestingly, hydrogen ions was found to have no effect on the electrical properties of the films, as the carrier concentration and Hall mobility values of Samples H2–H5 are similar to those of Samples B–E, respectively. The increase in the carrier concentration and Hall mobility of the annealed samples are only due to the

Table 2.4 Carrier concentration, Hall mobility and conductivity of different sample at 300 K

Sample name	Carrier concentration (cm^{-3})		Hall mobility (cm^2/V s)	
	Unimplanted	Implanted	Unimplanted	Implanted
As-deposited	1.01×10^{17}		16.1	
As-implanted	NA	4.5×10^{16}	NA	8.86
Annealed at 750 °C	1.13×10^{17}	1.24×10^{17}	18.3	33.9
Annealed at 800 °C	3.35×10^{17}	2.37×10^{17}	23.2	21.9
Annealed at 850 °C	5.75×10^{17}	8.85×10^{17}	60.6	54.4
Annealed at 900 °C	1.07×10^{18}	1.14×10^{18}	67.5	61.3

Reprinted from S. Nagar et al. "Increased photoluminescence of hydrogen-implanted ZnO thin films deposited using pulsed laser deposition technique", *Journal of Luminescence*, Vol. 15, pp. 307–311, 2014; with permission from Elsevier

reduction of the intrinsic defects of the films due to heat treatment. However, in all cases, the films demonstrated n-type behaviour which was obvious as hydrogen acts as donor in ZnO thin films [110–113].

2.4.1.3 Optical Properties

Low-temperature PL measurements of the obtained samples were executed in order to determine the optical properties of the implanted ZnO thin films (Fig. 2.20). A dominant donor-bound exciton (D°X) peak around 3.36 eV and a shoulder around 3.37 eV corresponding to the FX peak were observed for all the samples, which further confirms the n-type behaviour of the samples. The peaks corresponding to the extended defects due to implantation and the second LO-phonon replica of the D°X peak around 3.33 and 3.22 eV, respectively, were also observed in Samples H2–H6 [72, 104, 105]. Deep-level defect peaks around 2.5 eV, corresponding to defects such as Zn_i, V_O and implantation-related defects, were also observed for all the samples. However, the intensity of this peak is substantially weak when compared with the NBE peak intensity. This suggests that the ZnO films are of very high quality.

Temperature-dependent PL spectrum of different samples was performed to calculate the thermal activation energy of the samples using Eq. 2.3 (Table 2.5). Figure 2.21 shows the integrated PL peak intensity of the D°X peak for various samples with respect to temperature.

$$I = I_0 * (1 + \exp(-E_A/kT)) \tag{2.3}$$

where I and I_0 are the integrated PL peak intensity at temperature T K and 8 K, E_A is the activation energy and k is the Boltzmann constant.

The activation energy was found to decrease for Sample H1 due to the increased number of defects caused by implantation. On annealing the samples, the activation energy increases marginally with temperature. However, the activation energy is

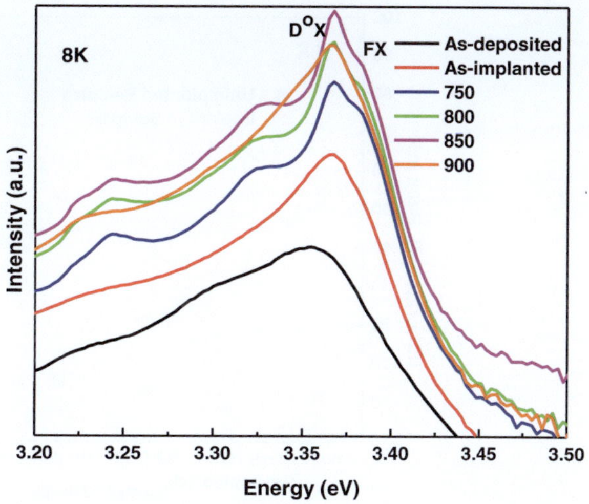

Fig. 2.20 PL spectra of different samples taken at 8 K (Reprinted from S. Nagar et al. "Increased photoluminescence of hydrogen-implanted ZnO thin films deposited using pulsed laser deposition technique", *Journal of Luminescence*, Vol. 15, pp. 307–311, 2014; with permission from Elsevier)

Table 2.5 Thermal activation energy for different samples

Sample name	Thermal activation energy (meV)
As-deposited	66.7
As-implanted	26.5
Annealed at 750 °C	36.98
Annealed at 800 °C	38.6
Annealed at 850 °C	40.5
Annealed at 900 °C	41

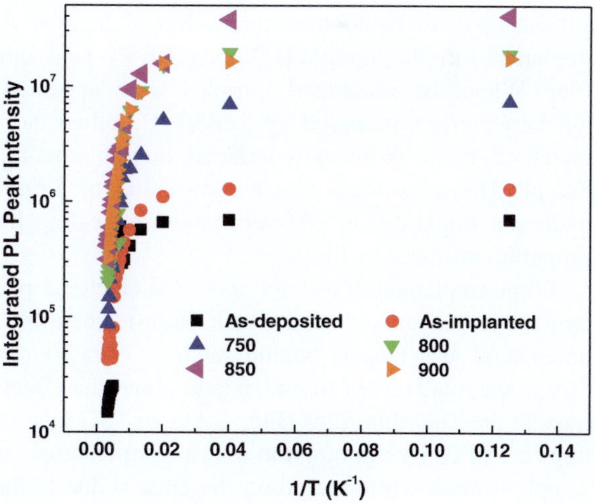

Fig. 2.21 Integrated PL peak intensity with respect to temperature for different samples

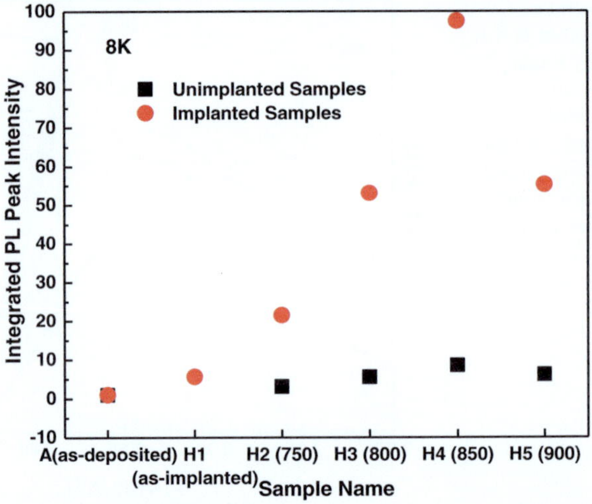

Fig. 2.22 Photoluminescence (PL) peak intensity at 8 K for unimplanted and implanted samples for the D°X peak (Reprinted from S. Nagar et al. "Increased photoluminescence of hydrogen-implanted ZnO thin films deposited using pulsed laser deposition technique", *Journal of Luminescence*, Vol. 15, pp. 307–311, 2014; with permission from Elsevier)

still lower than that of Sample A. The formation of shallow donor levels owing to hydrogen implantation leads to lower activation energy in Samples H2−H6 [110–113].

Figure 2.22 illustrates the NBE PL peak intensity for the implanted and unimplanted samples. Notably, the figure shows an obvious increase in the NBE PL peak intensity for the unimplanted sample, with increase in annealing temperature, due to the reduction of defects. A sample annealed at 850 °C demonstrated the highest enhancement of up to nine times that of Sample A. On the other hand, for the implanted sample (Sample H1), the NBE PL peak intensity improved by a factor of four. When the implanted samples were annealed at higher temperatures, the intensity further increased up to 850 °C. When the annealing temperature further increased, the peak intensity reduced, though remaining quite high. The intensity of Sample H4 is as high as 100 times that of Sample A. This result shows that hydrogen implantation and subsequent annealing had some effect on the optical properties of the thin films.

While implantation did not affect the electrical properties of the thin films, their optical properties revealed much improvement in terms of optical output. To understand the reason behind such a behaviour, thermally stimulated current (TSC) was undertaken to understand about the effect of implantation on the defect traps in the ZnO thin films (Fig. 2.23). In TSC measurements, the electron and hole traps are filled by illumination at low temperatures and by subsequent warming the sample in dark while recording the current due to the thermal emission of carriers

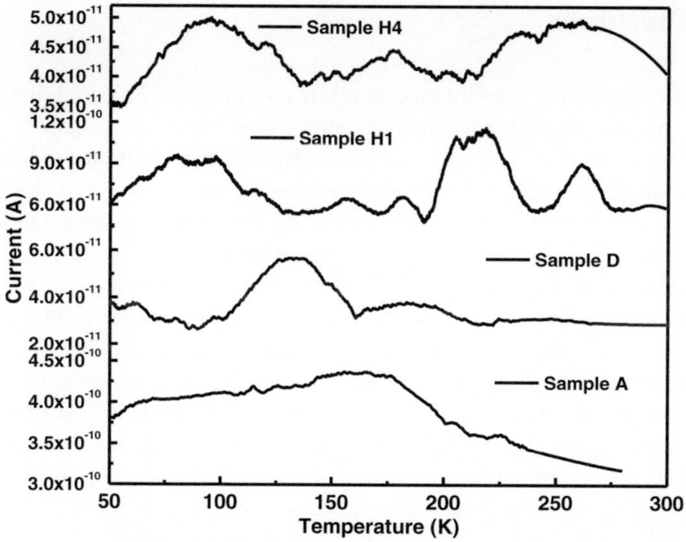

Fig. 2.23 Thermally stimulated current spectra of Samples A, D, H1 and H4 (Reprinted from S. Nagar et al. "Increased photoluminescence of hydrogen-implanted ZnO thin films deposited using pulsed laser deposition technique", *Journal of Luminescence*, Vol. 15, pp. 307–311, 2014; with permission from Elsevier)

from the traps [114]. All the samples were first cooled down from room temperature to 8 K, during which the dark current was measured with respect to temperature.

Illumination, for trap filling at 8 K, was performed with a xenon lamp for 30 min. The TSC current was then subsequently measured while warming up at a rate of 5 K/min under a bias of 2 V. A TSC peak around 162 K is observed for the as-deposited sample, which is resolved into two peaks upon annealing at 132 and 185 K. Although TSC measurements cannot reveal the exact nature of the trap, it is believed that these peaks are due to V_O, as they have activation energy in the range of 0.25–0.33 eV calculated from Eq. 2.4 [114–117].

$$E_T = kT_m \ln(T_m^4/\beta) \tag{2.4}$$

where E_T is the activation energy of a related trap, T_m is the peak temperature for the particular trap and β is the heating rate. Upon implantation and subsequent annealing, three other peaks at 87, 215 and 261 K are observed along with relatively small peaks around 160 and 185 K. The exact origins of these peaks are still unknown, but the hydrogen implantation provides clues to the origins of these traps. As the electrical properties of the films do not change even after annealing, it may be possible that these traps are electrically inactive. However, during PL measurements, the trapped electrons are released due to laser illumination, giving rise to increased PL intensity in these peaks. Such an enhanced optical luminescence in ZnO films can be utilised to fabricate devices with very high optical efficiency.

2.4.2 Conclusion

The effects of hydrogen implantation at a low energy of 50 keV and of subsequent annealing on the PLD-deposited ZnO thin films were investigated. A strong c-axis-oriented ⟨002⟩ ZnO films have been deposited in each case. However, the strain developed in the films due to implantation caused the peaks to shift to higher 2θ angles. Van der Pauw Hall measurements did not show any change in the carrier concentration and Hall mobility of the implanted and annealed samples, compared to as-annealed samples. While the PL results showed a dominant D°X for all the samples, a significant increase in the PL emission (×4 and ×100 times) was observed for the implanted sample and sample annealed at 850 °C. The formation of hydrogen-related traps due to implantation has caused an increased PL emission from the sample. To the best of our knowledge, this report is one of the first in which an improvement in the luminescence intensity has been observed. Such an enhancement in the optical luminescence of the ZnO thin films may be helpful in fabricating highly efficient ZnO-based devices.

Chapter 3
Implantation Studies

Abstract Ion implantation is a key tool that has developed in recent years for the fabrication of devices. It has the added advantage of providing lateral selectivity, appropriate doping area and fluence control along with localised doping. Ion implantation studies have been performed with the goal of doping ZnO with group I and group V elements to make p-type films. The Low-Energy Accelerator Facility at BARC, Mumbai, was used to implant lithium and phosphorus ions. Lithium ions were found to occupy deep acceptor states in ZnO and thus acting as donors in the films. Hence, no acceptor peaks were observed in the films implanted with lithium. On the other hand, when the films were implanted with phosphorus ions, an increase in the acceptor peaks was observed for the implanted and annealed films. However, the films were still n-type in nature. Plasma immersion ion implantation technique was used to implant the sputter-deposited ZnO thin films with phosphorus and nitrogen ions followed by subsequent annealing to remove any implantation-related defects. In both cases, a strong acceptor-bound-exciton (A°X) peak was observed around 3.35 eV, from PL measurements, in the thin films annealed at temperatures of 900–1000 °C confirming that the films have converted into p-type. The acceptor activation energy was calculated to be 125 and 118 eV for phosphorus- and nitrogen-implanted samples, respectively. This proves that phosphorus ion and nitrogen ion act as shallow acceptor levels in ZnO giving rise to p-type behaviour. PL measurements for the p-type samples were performed over a period of time to check the reliability of the p-type films. For the phosphorus-implanted samples, the p-type was found to exist up to a period of 6 months, although the PL intensity was found to reduce drastically. For the

Part of the chapter have been reprinted from the following:[S. Nagar et al. "Structural, electrical and optical characteristics of lithium-implanted ZnO thin films", Superlattices and Microstructures, Vol. 53, Pg 164–172, 2013][S. Nagar et al. "Effect of phosphorus irradiation in the structural, electrical and optical characteristics of ZnO thin films", Journal of Luminescence, Vol. 132, Pg 1089–1094, 2012][S. Nagar et al. "Evidence of strong acceptor peaks in ZnO thin films doped with phosphorus by plasma immersion ion implantation technique", Journal of Luminescence, Vol. 137, Pg 55–58, 2013][S. Nagar et al. "Realization of reliable p-type ZnO thin films by nitrogen implantation using plasma immersion ion implantation", Superlattices and Microstructures, Vol. 75, Pg 9–16, 2014]

S. Nagar and S. Chakrabarti, *Optimisation of ZnO Thin Films*,
DOI 10.1007/978-981-10-0809-2_3

nitrogen-implanted samples, the p-type was seen even after a period of 9 months and even the PL peak intensity was similar to the freshly made sample.

Keywords Low-Energy Accelerator Facility · Plasma immersion ion implantation · Annealing · Reliability

3.1 Introduction

Ion implantation is one of the key tools in the field of device fabrication. It has the added advantage of providing lateral selectivity, appropriate doping area and fluence control along with localised doping. Ion implantation technique was used to dope elements in ZnO thin films with group I and group V elements.

3.2 Conventional Ion Implantation Studies

CII was performed on ZnO film, deposited by the PLD technique. The Low-Energy Accelerator Facility (LEAF) at BARC, Mumbai, was used for the purpose. We doped lithium (group I) and phosphorus (group V) ions with the intention of achieving p-type ZnO films.

3.2.1 Deposition of ZnO

ZnO thin films have been deposited using PLD system over ⟨001⟩ sapphire substrates. A 99.999% pure ZnO target (1 inch in diameter) was used for ablation via a KrF excimer laser (248 nm, 20 ns pulse). The energy density used for the ablation was fixed at 1.8 J/cm^2. The substrate was first degreased properly using trichloroethylene (TCE), acetone and isopropyl alcohol (IPA) for 2 min each prior to deposition at 400 °C. The target-to-substrate distance was kept at 5 cm. The deposition was performed in a high-vacuum chamber under a background oxygen pressure of 75 mTorr for 40 min. The thickness of the films thus obtained was measured to be around 500 nm using α-profiler. Prior to deposition, the chamber was kept in a high vacuum of 3×10^{-5} mTorr. Thus, Sample A1 was obtained.

3.2.2 Lithium Implantation Studies

3.2.2.1 Experimental Details

Li ion implantation was performed over the deposited films at low energies of 40 and 50 keV and a constant dose of 5×10^{13} ions/cm^2 at room temperature, in order to obtain samples B1 and C1, respectively, using LEAF at BARC, Mumbai. The projected range of 165 and 207 nm, respectively, was estimated using TRIM calculations. Samples B1 and C1 were subjected to rapid thermal annealing (AS ONE 150, Annealsys) at 650–750 °C, in Ar environment for 30 s to yield Samples B2, B3 (implanted at 40 keV) and C2, C3 (implanted at 50 keV), respectively.

3.2.2.2 Results and Discussion

Structural Properties

Figures 3.1 and 3.2 depict the XRD pattern of the obtained samples. The presence of ⟨002⟩ peak in all the samples confirms growth of highly c-axis-oriented films. The drastically diminished peak intensity for Samples B1 and C1 (compared to A1) implies deterioration of film crystallinity due to inclusion of Li ions, deterioration which is subsequently regained on annealing. This phenomenon can be further verified by looking at the FWHM values of the samples (Table 3.1): Samples B2, B3 and C2, C3 show lower values, thus confirming the reduced defects and greater crystallinity.

AFM study was performed to study the surface morphology of the obtained samples (Figs. 3.3 and 3.4). Figures 3.3b and 3.4b show better grain formation in Samples B1 and C1 as compared to that in A1 (as shown in Figs. 3.3a, 3.4a).

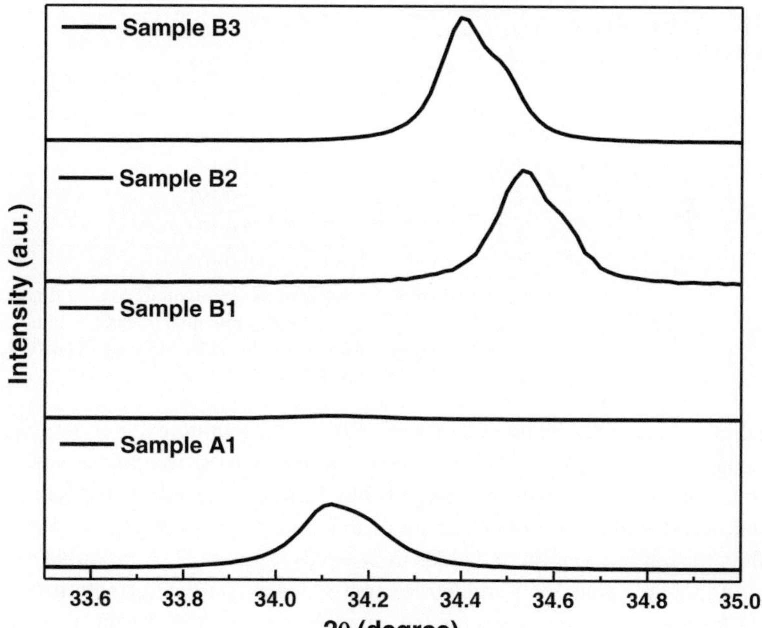

Fig. 3.1 XRD pattern for films obtained after implantation at 40 keV showing the ⟨002⟩ ZnO peaks

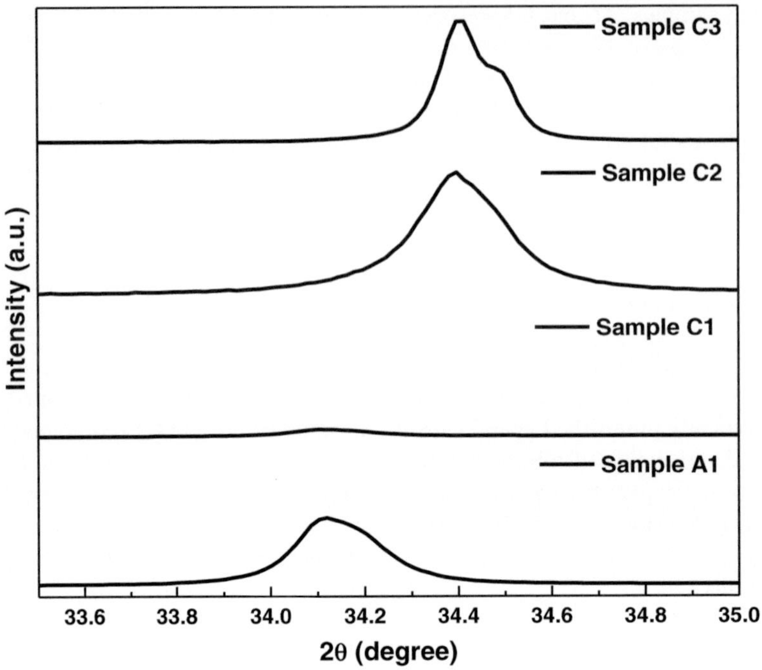

Fig. 3.2 XRD pattern for films obtained after implantation at 50 keV showing the ⟨002⟩ ZnO peaks

Table 3.1 FWHM values of the various films

Sample name	FWHM value (°)
A1	0.217
B1	0.217
B2	0.178
B3	0.171
C1	0.191
C2	0.163
C3	0.102

A possible reason might be the generation of heat in the sample due to implantation, which causes an increase in the root mean square (RMS) roughness value, measured over an area of 1 μm × 1 μm (Table 3.2). As the implanted samples are subjected to annealing at 650 °C, some of the smaller grains merge to form larger ones, thus raising the roughness values in Samples B2 and C2. A probable reason of higher roughness value for Samples B3 and C3 may be the formation of micro-voids, resulting from the merging of smaller grains into larger ones due to high-temperature annealing.

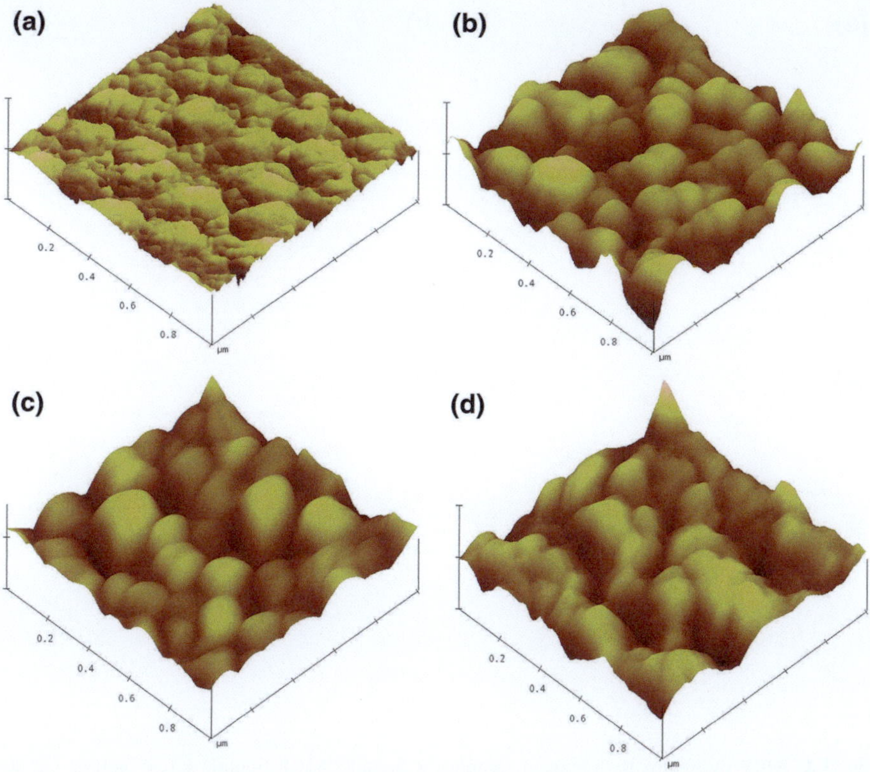

Fig. 3.3 AFM figures of the obtained samples. **a** Sample A1, **b** Sample B1, **c** Sample B2 and **d** Sample B3

Surface morphology of the obtained samples was assessed from SEM images, as shown in Figs. 3.5 and 3.6. Better grain formation was observed in Samples B1 and C1 (Figs. 3.5b and 3.6b), as compared to the Sample A1 (Figs. 3.5a and 3.6a). A possible cause of better grain formation might be related to the generation of heat in those samples during implantation. Formation of microvoids can also be observed in these samples, which may be due to the increased defects in those samples.

Electrical Properties

The effects of implantation on the film's electrical properties were surveyed by carrying out temperature-dependent Van der Pauw Hall measurements of the samples in the range of 80–300 K (Figs. 3.7 and 3.8). Carrier concentration, Hall mobility and resistivity of various samples at room temperature are shown in Table 3.3. The implanted samples exhibited the expected n-type characteristics over the entire temperature range due to the presence of implantation-induced defects, and they showed an increase in carrier concentration in B1 and a reduced carrier

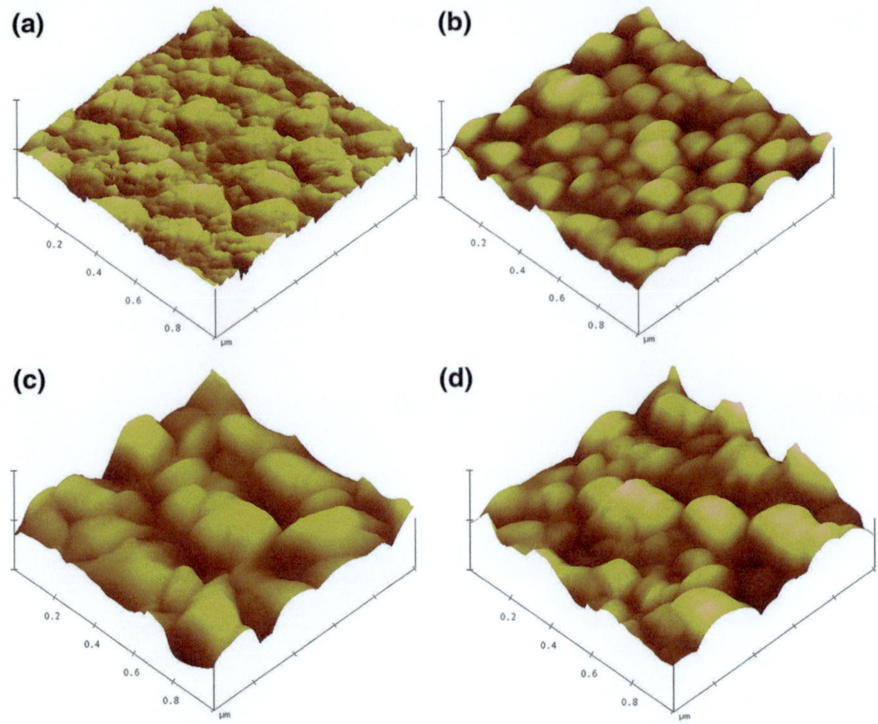

Fig. 3.4 AFM figures of the obtained samples. **a** Sample A1, **b** Sample C1, **c** Sample C2 and **d** Sample C3

Table 3.2 Root mean square (RMS) roughness values of the samples

Sample name	RMS roughness (nm)
A1	7.09
B1	8.12
B2	12.37
B3	13.47
C1	8.11
C2	12.87
C3	13.47

concentration in C1 as compared to A1. The reduction in concentration in C1 may be due to fewer defects and denser grains as seen from AFM and SEM images. The implanted samples were subjected to rapid thermal process at 650–750 °C, not only to reduce the defects but also to activate the Li ions to occupy Zn interstitial sites and lead to p-type films. However, the samples continued to show n-type characteristics. At the same time, significant changes in the carrier concentration, resistivity and Hall mobility were noticed in these films. For the samples annealed at 650 °C (Samples B2 and C2), the carrier concentration and resistivity are found to reduce. However,

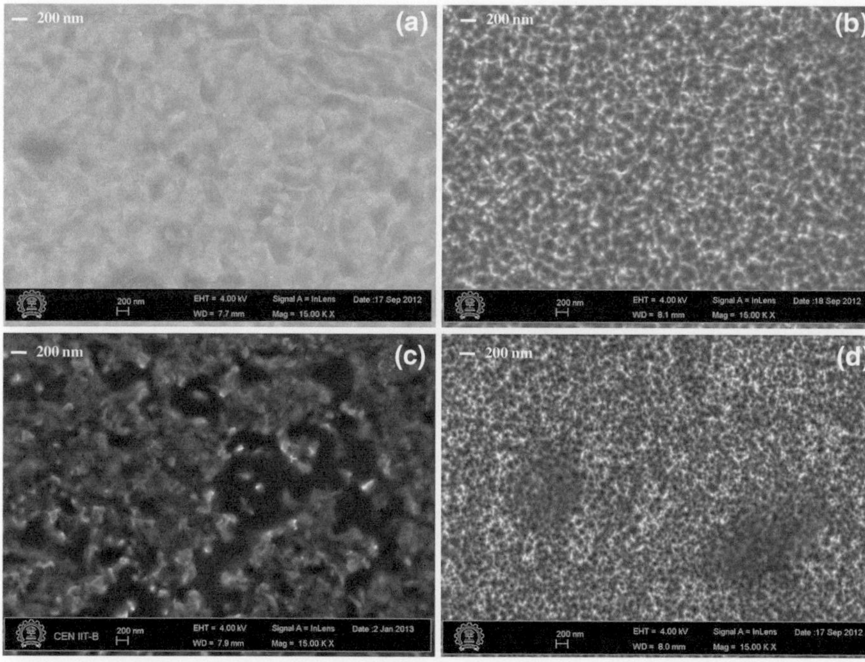

Fig. 3.5 SEM images. **a** Sample A1, **b** Sample B1, **c** Sample B2 and **d** Sample B3

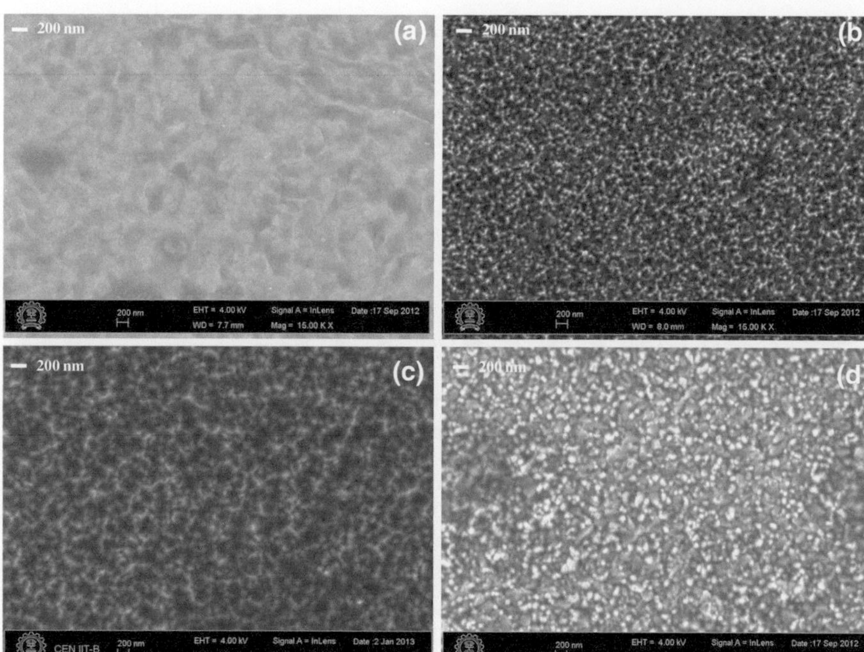

Fig. 3.6 SEM images. **a** Sample A1, **b** Sample C1, **c** Sample C2 and **d** Sample C3

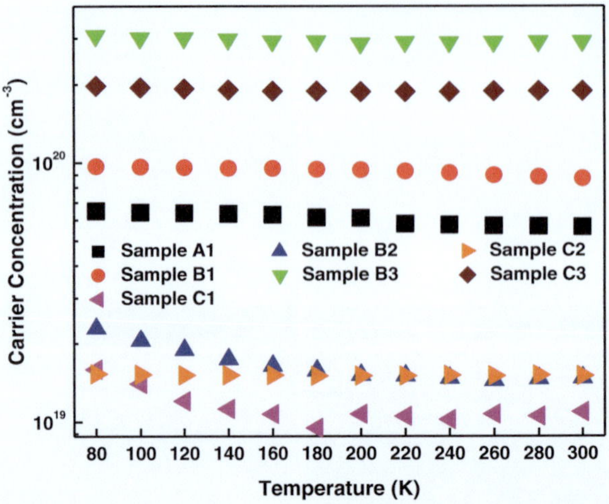

Fig. 3.7 Temperature-dependent carrier concentration for the obtained samples (Reprinted from S. Nagar et al. "Structural, electrical and optical characteristics of lithium-implanted ZnO thin films", *Superlattices and Microstructures*, Vol. 53, pp. 164–172, 2013; with permission from Elsevier)

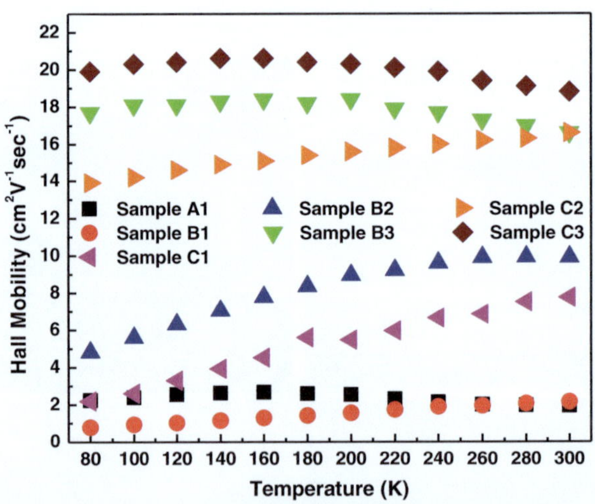

Fig. 3.8 Temperature-dependent Hall mobility for the obtained samples (Reprinted from S. Nagar et al. "Structural, electrical and optical characteristics of lithium-implanted ZnO thin films", *Superlattices and Microstructures*, Vol. 53, pp. 164–172, 2013; with permission from Elsevier)

Table 3.3 Carrier concentration, Hall mobility and resistivity values of the obtained samples taken at room temperature

Sample name	Carrier concentration (cm^{-3})	Hall mobility (cm^2/V s)	Resistivity (Ω cm)
A1	5.64×10^{19}	1.94	5.35×10^{-2}
B1	8.68×10^{19}	2.14	3.4×10^{-2}
B2	1.13×10^{19}	35.2	1.57×10^{-2}
B3	2.9×10^{20}	16.9	1.27×10^{-2}
C1	2.1×10^{19}	7.16	4.15×10^{-2}
C2	1.51×10^{19}	21	1.96×10^{-2}
C3	1.95×10^{20}	20.2	1.59×10^{-3}

for Samples B3 and C3, the carrier concentration increases, while resistivity decreases. The aggravation of vacancy clusters due to implantation, which grows into microvoids, might be a probable reason for the increase in concentration. These microvoids are easier to recover at lower temperatures, thus explaining lower concentration for Samples A2 and B2 [68]. Moreover, Lopatiuk et al. have also reported that the introduction of lithium ions in ZnO films tends to introduce deep electron traps [118]. The combination of deep electron traps and microvoids may have led to such an observation. Moreover, for all the annealed samples, the Hall mobility increases, probably due to minimised implantation-related defects.

Optical Properties

Low-temperature (8 K) PL spectra were analysed to assess the change in optical properties due to implantation (Figs. 3.9 and 3.10). A prominent NBE around 3.36 eV, corresponding to the donor-bound-exciton (D°X) peak, was observed in all implanted and annealed samples, confirming the n-type conductivity of the films, as shown by the Hall measurement results [72, 104]. It is interesting to note that no deep-level peaks were observed for any of the samples. This may be due to the small ionic radii of Li ions, which would not markedly affect the crystal structure, although implantation-related defects do occur, as seen from SEM images. Another peak is observed around 3.33 eV, which is attributed to the exciton bound to an extended defect, as reported by Wagner et al. [105]. The intensity of this peak reduces for Samples B1, B3, C1 and C3 compared with A1, perhaps due to the better grain formation in these samples. But for Sample B2, this peak increases, while the peak was not observed in Sample C2. The intensity of this peak seems to increase for Samples B3 and C3 when compared with B1 and C1. A possible reason might be the increased microvoid formation in the samples as observed by SEM

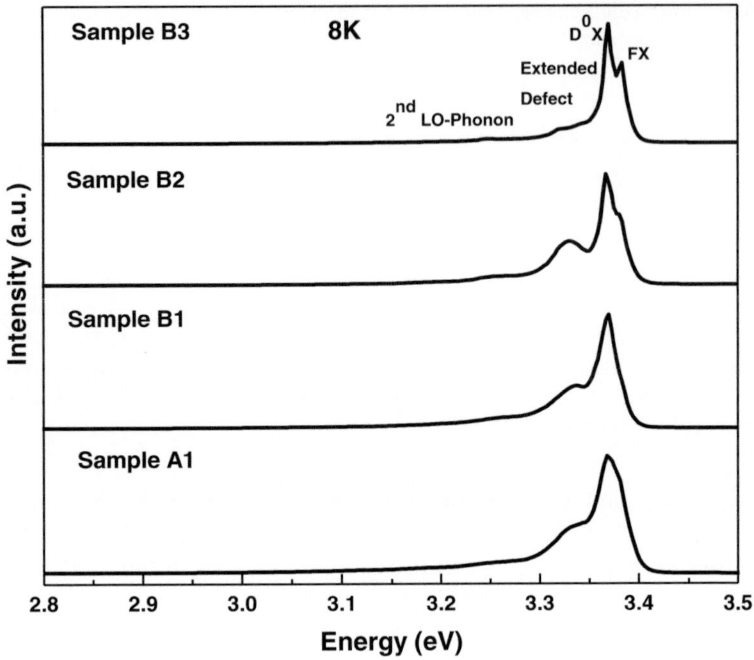

Fig. 3.9 PL spectra of the samples, obtained after implantation at 40 keV, taken at 8 K. No deep-level peaks in the samples are observed

images. A peak is observed around 3.22 eV, which is attributed to the second LO-phonon replica because:

$$3.365 - 0.072 - 0.072 = 3.221 \tag{3.1}$$

where 3.365 eV is the energy corresponding to the D°X peak and 0.072 eV is the energy of the LO phonon. The absence of the first LO-phonon replica may be because it is hidden in the background of the exciton-bound defect peak around 3.33 eV. A high-intensity LO-phonon peak was observed in Sample C2 and may be a reason for the absence of the extended defect peak around 3.33 eV. For Samples B3 and C3, FX peak around 3.38 eV was also present. Due to the increased carrier concentration (as determined by Hall measurements) in these samples, some excitonic transitions might have occurred, which would explain the observed FX peak. However, no acceptor peak was visible in any of the samples.

The temperature-dependent PL spectra of all samples are shown in Figs. 3.11 and 3.12. The prominent D°X peak at lower temperatures started to decrease with temperature, while the FX peak started to dominate at higher temperatures. The enhanced thermal dissociation of bound excitons into free excitons causes this kind

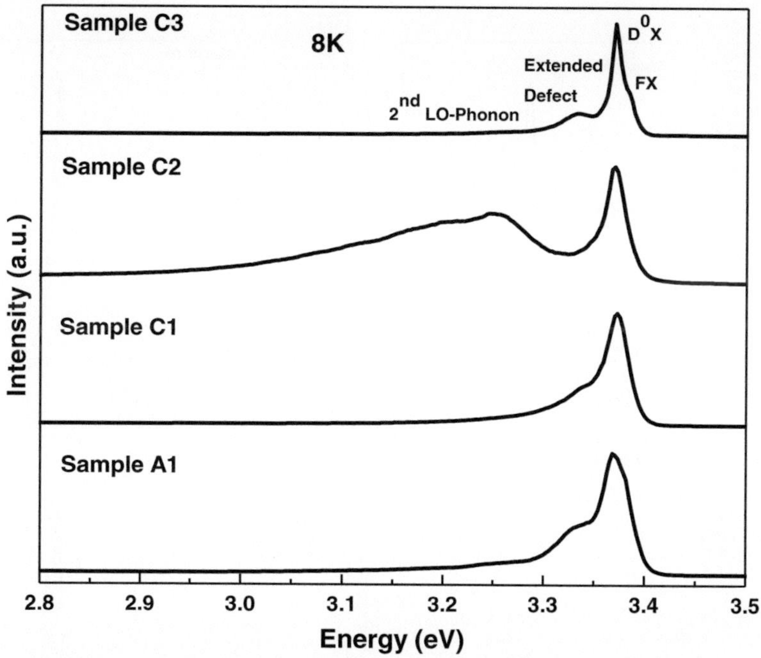

Fig. 3.10 PL spectra of the samples, obtained after implantation at 50 keV, taken at 8 K. No deep-level peaks in the samples are observed

of observation [108]. To depict the optical quality of the films, the integrated PL intensity of the obtained samples was plotted against the inverse of the temperature (Figs. 3.13 and 3.14). The results indicate that both implanted samples have better optical behaviour than do the as-deposited samples, which may be due to the denser grain formation in the implanted samples (Figs. 3.3b, 3.4b, 3.5b and 3.6b). As shown in Fig. 3.13, Sample B2 provides the best luminescence among all the samples. Figure 3.14 depicts samples implanted with 50 keV of energy. In this case, Sample C2 shows the least luminescence, while Sample C3 has the highest. The presence of microvoids in C2 may have caused the reduction in its optical quality, whereas the presence of denser grains in C3 led to its higher optical quality.

Quantitative fitting of different samples was done, using Figs. 3.13 and 3.14, to calculate the thermal activation energy using Eq. 2.3. The results have been tabulated in Table 3.4. The increase of defects and electron traps leads to a reduction in activation energy for B1 and C1. Upon annealing of the implanted samples, the activation energy increased to a high value of 147.28, 176.54, 223.04 and 236.59 meV for Samples B2, B3, C2 and C3, respectively. The reduced strain and defects in the films due to annealing lead to this increase in activation energy. Moreover, it has been shown by Wardle et al. [119] and Lopatiuk et al. [118] that

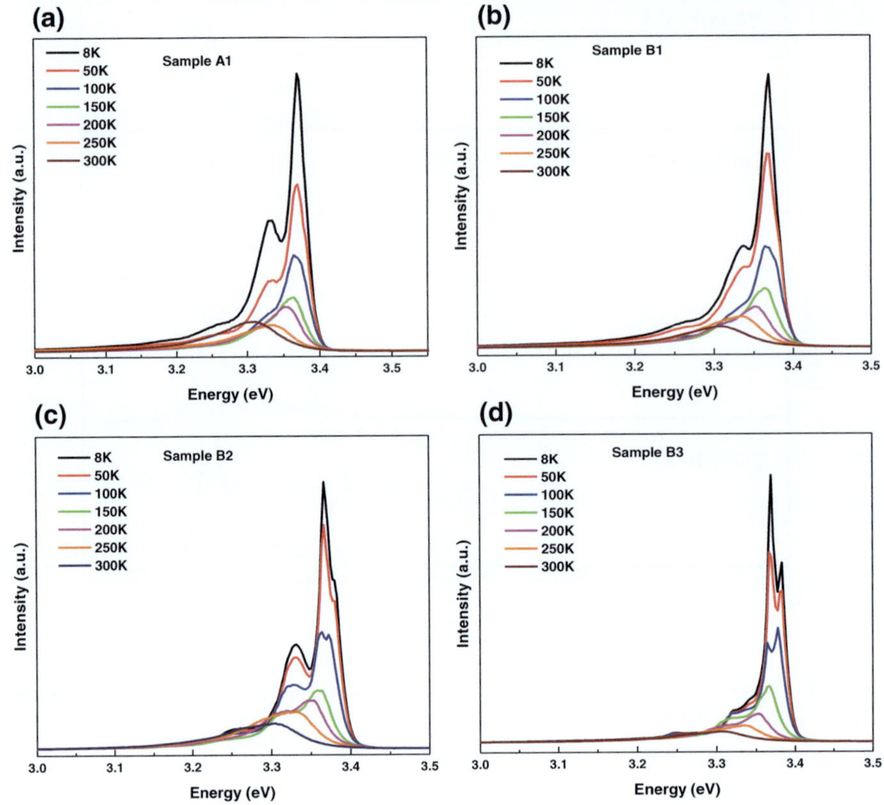

Fig. 3.11 Temperature-dependent PL spectra for Samples A1, B1, B2 and B3

the high values in these samples are due to the deep Li acceptors, which cause trapping of non-equilibrium electrons.

3.2.2.3 Conclusion

Li ion implantation was performed on ZnO films at low energies (40 and 50 keV) with the intention to make p-type films, and a detailed study on the structural, electrical and optical properties was carried out. Subsequent annealing of the samples was performed at 650–750 °C. XRD results confirmed the deposition of highly c-axis-oriented ⟨002⟩ ZnO films. Hall measurement results revealed increased electron carrier concentration for samples annealed at 750 °C. Although no acceptor peaks were visible in the PL spectra of the implanted samples, significant changes with respect to the defect have been observed. The implantation of Li ions has some significant changes in the electrical and optical quality of the ZnO

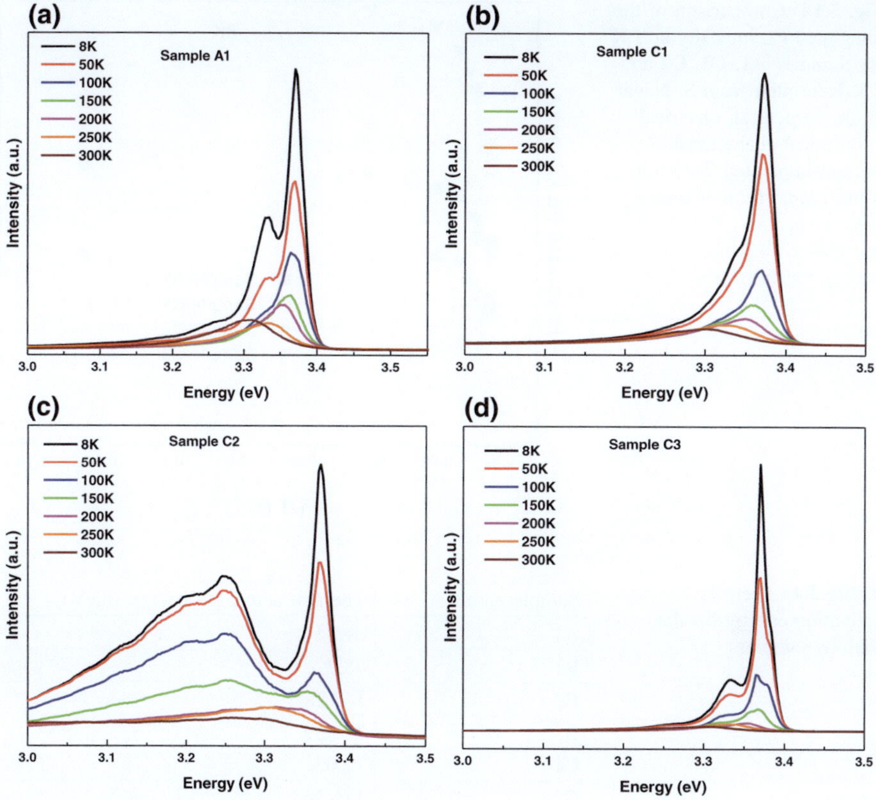

Fig. 3.12 Temperature-dependent PL spectra for Samples A1, C1, C2 and C3

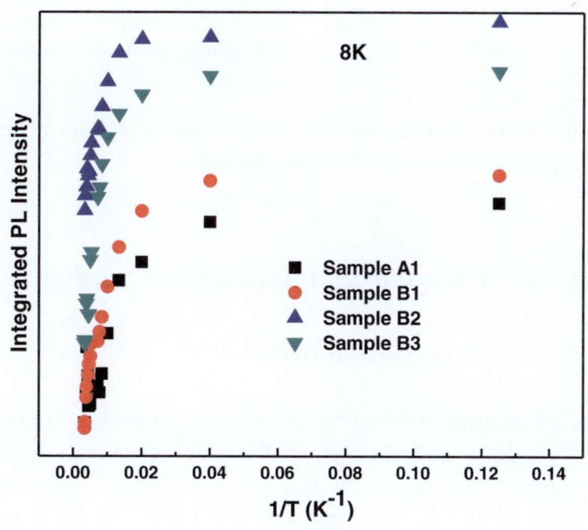

Fig. 3.13 Comparison of the integrated PL intensity at 8 K for Samples A1, B1, B2 and B3 (Reprinted from S. Nagar et al. "Structural, electrical and optical characteristics of lithium-implanted ZnO thin films", *Superlattices and Microstructures*, Vol. 53, pp. 164–172, 2013; with permission from Elsevier)

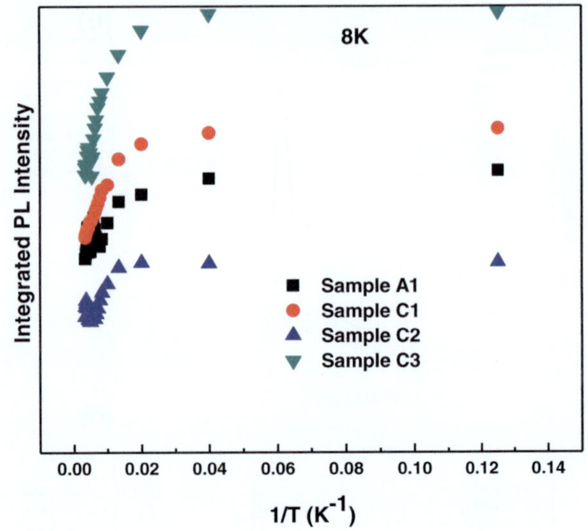

Fig. 3.14 Comparison of the integrated PL intensity at 8 K for Samples A1, C1, C2 and C3 (Reprinted from S. Nagar et al. "Structural, electrical and optical characteristics of lithium-implanted ZnO thin films", *Superlattices and Microstructures*, Vol. 53, pp. 164–172, 2013; with permission from Elsevier)

Table 3.4 Thermal activation energy for the various samples

Sample name	Thermal activation energy (meV)
A1	89.84
B1	68.3
B2	147.28
B3	236.59
C1	72.07
C2	176.54
C3	223.04

Reprinted from S. Nagar et al. "Structural, electrical and optical characteristics of lithium-implanted ZnO thin films", *Superlattices and Microstructures*, Vol. 53, pp. 164–172, 2013; with permission from Elsevier

thin films caused when Li forms deep acceptors in the ZnO films, thus leading to trapping of non-equilibrium electrons.

3.2.3 *Phosphorus Implantation Studies*

3.2.3.1 Experimental Details

Implantation of phosphorus ions was carried out using a beam accelerating energy of 50 keV and dosage of 8×10^{14} ions/cm^2 (Sample D1). LEAF at BARC, Mumbai, was used for this purpose. TRIM calculations measured the projected range of 917 Å. After implantation, rapid thermal annealing process was carried out in Ar ambient at temperatures of 650–750 °C to obtain Samples D2 and D3, respectively. One of the samples was also annealed at 750 °C in oxygen environment to visualise the effect of various background gases on the implanted films (Sample D4).

3.2.3.2 Results and Discussion

Structural Properties

XRD studies reveal deposition of highly c-axis-oriented ⟨002⟩ films (Fig. 3.15). The presence of only (001) family peaks certifies preferential growth of ZnO in the ⟨001⟩ direction due to its lowest surface free energy. The FWHM of the Sample D1 was higher than in A1, which is to be expected, as implantation causes defects in the film. While Sample D2 did show slight improvement in the FWHM as compared to Sample D1, the FWHM value was considerably lower in Samples D3 and D4 indicating that the defects caused by implantation have reduced when annealed at higher temperatures. It is also interesting to note that for Samples D1, D2 and D3, only the (001) family of peaks is visible, while Sample D4 also showed a small peak at 31.5° corresponding to ⟨100⟩ direction clearly representing the

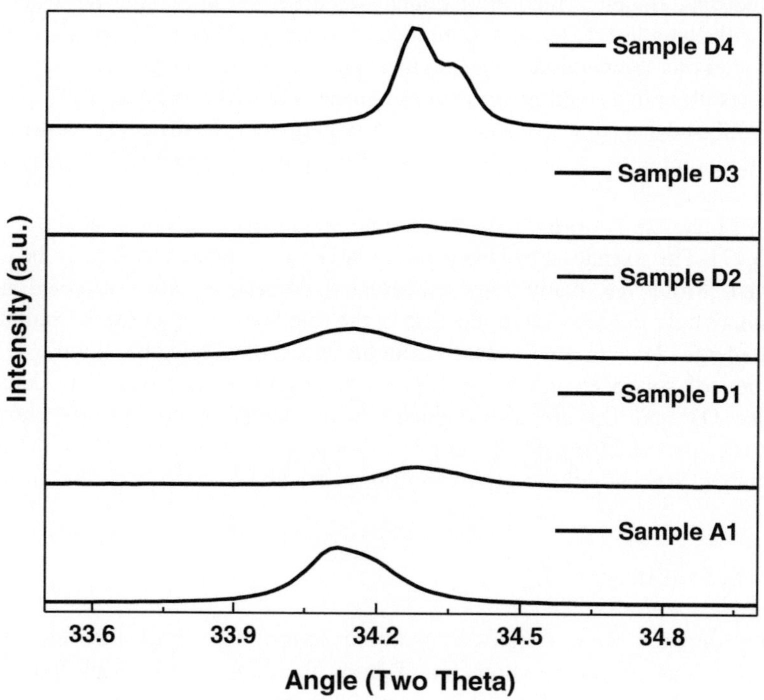

Fig. 3.15 XRD patterns for the obtained films showing the ⟨002⟩ and ⟨004⟩ ZnO peaks

Table 3.5 FWHM values of different films

Sample name	FWHM value (°)
A1	0.217
D1	0.257
D2	0.224
D3	0.183
D4	0.165

Reprinted from S. Nagar et al. "Effect of phosphorus irradiation in the structural, electrical and optical characteristics of ZnO thin films", *Journal of Luminescence*, Vol. 132, pp. 1089–1094, 2012; with permission from Elsevier

polycrystalline nature of the film. Thus, the effect of background gas environment during annealing can be easily identified. Table 3.5 shows the FWHM values of different samples.

Atomic force microscopy was performed to understand the morphological changes observed in the samples with implantation and annealing in different environments (Fig. 3.16). Very little grain formation can be detected for Sample A1 when compared to Samples D1, D2 and D3. The internal heating of the samples as a result of implantation leads to substantial grain formation. This tends to increase the RMS roughness for Sample D1 when compared to A1, measured over an area of 1 μm × 1 μm (Table 3.6). The RMS roughness values for Samples D2 and D3 were reduced. Better formation of grains was observed in Sample D2 when compared with Sample D1, leading to reduced roughness. When the sample is further heated at higher temperatures, then some small grains merge together to form larger grains, resulting in a slight increase in roughness. However, in Sample D4, the same grains diffused, causing the roughness to continue to decrease. The diffusion of some grains might be a probable reason for the polycrystallinity of the film, as has been discussed above.

SEM imaging performed on the samples further validates the above claim (Fig. 3.17). The formation of better grains in D1 as compared to A1 can be clearly observed. Moreover, many implanted-related defects are also observed in D1, which may have contributed to increase in the roughness of the film. In Sample D2, grains merge, leading to increased roughness of the films. In Sample D3, the formation of spherical grains can be clearly observed. Moreover, the defects in Samples D3 and D4 are also reduced when compared to D1. As shown in Fig. 3.16e, the SEM image of Sample D4 also proves that annealing in oxygen environment has caused the grains to diffuse, leading to polycrystalline films.

Electrical Properties

All the samples were subjected to room temperature Hall measurement by employing Van der Pauw geometry to study the effect of the implantation. The results, as tabulated in Table 3.7, clearly exhibit n-type behaviour for all the samples. A slight increase in the carrier concentration and consequent decrease in

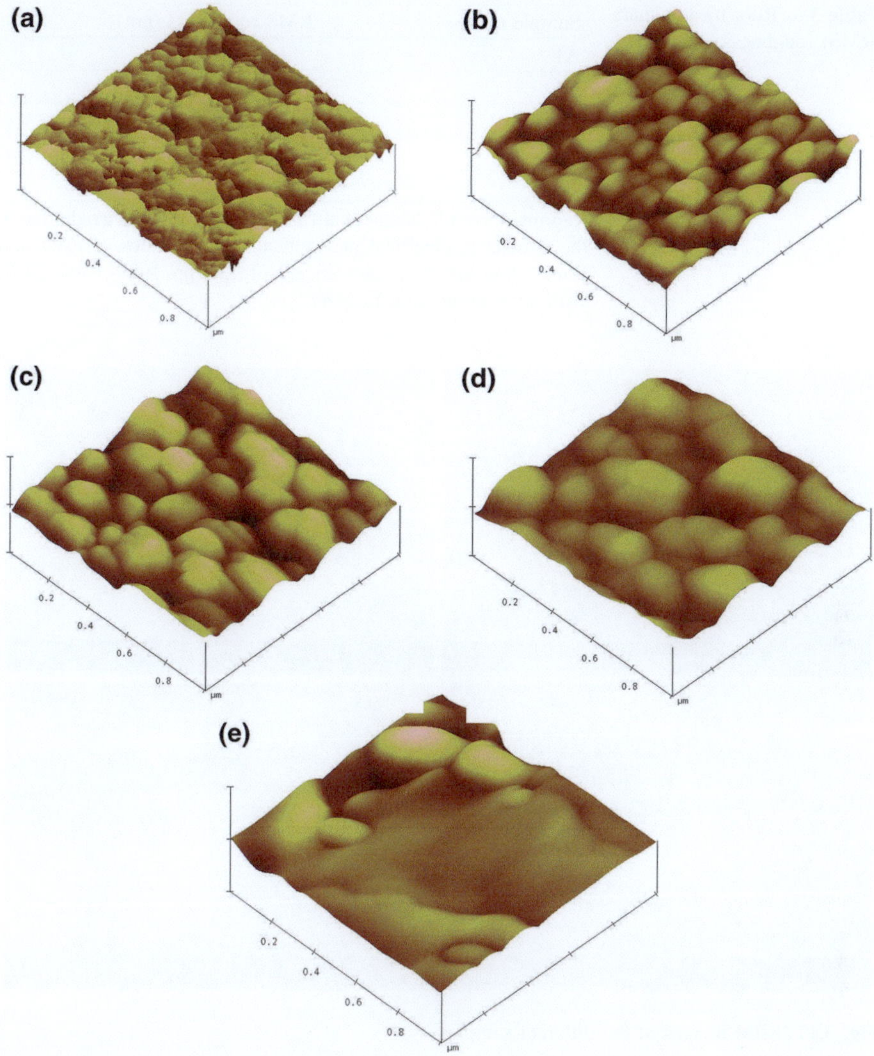

Fig. 3.16 AFM images of the obtained samples: **a** Sample A1, **b** Sample D1, **c** Sample D2, **d** Sample D3 and **e** Sample D4 (Figures (**a**–**d**) reprinted from S. Nagar et al. "Effect of phosphorus irradiation in the structural, electrical and optical characteristics of ZnO thin films", *Journal of Luminescence*, Vol. 132, pp. 1089–1094, 2012; with permission from Elsevier)

Hall mobility appear in Sample D1, which is to be expected due to the increase in the number of ions as a result of implantation. A marked reduction in the carrier concentration and subsequent increase in Hall mobility were observed for Sample D2, while in D3 the carrier concentration and Hall mobility increased. The improvement in mobility might be due to the reduction of defects in heat treatment. It can be clearly seen from SEM images that the grains in D2 are small and very

Table 3.6 Root mean square (RMS) roughness of samples

Sample name	RMS roughness (nm)
A1	7.09
D1	10.07
D2	8.66
D3	9.31
D4	4.9

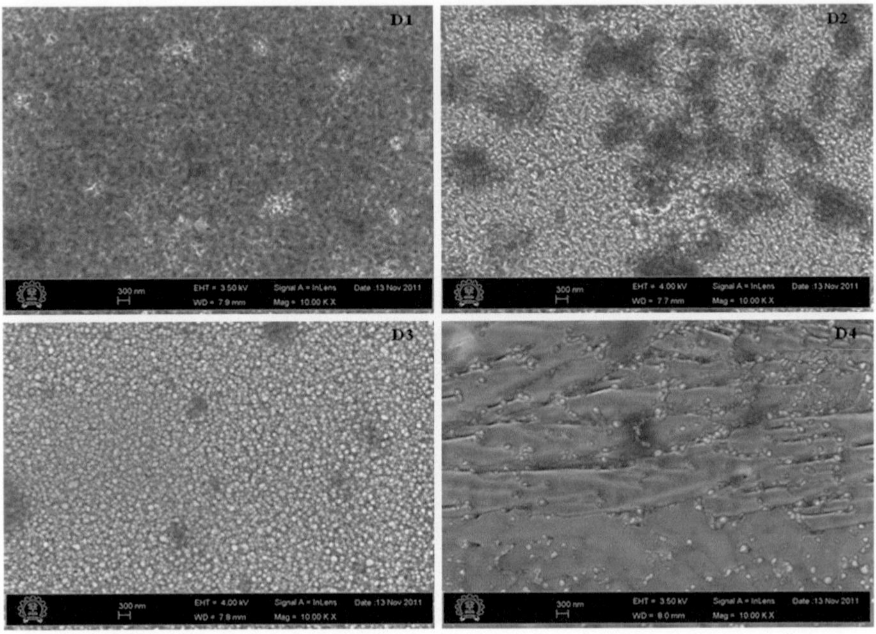

Fig. 3.17 SEM images of the different samples

closely packed, unlike in D3. As a result, the grain boundary scattering in D3 is much larger compared to that of D2, leading to higher carrier concentration values and lower Hall mobility values. The comparative smooth surface of Sample D4 with respect to other samples in combination with annealing effects has caused the reduction in carrier concentration and increase in the Hall mobility for the sample. The reduction in concentration and increase of mobility for most of the annealed samples prove that compensation of n-type character has taken place due to P-doping. Temperature-dependent Hall measurements were performed in all the samples to check whether lowering of temperature caused any type conversion in the samples (Figs. 3.18 and 3.19). However, no such observation was found, and samples continuously depicted n-type carriers in the whole temperature range.

Table 3.7 Carrier concentration and Hall mobility of various samples at room temperature

Sample name	Type of conductivity	Carrier concentration (cm^{-3})	Mobility ($cm^2\ V^{-1}\ s^{-1}$)
A1	n-type	4.5×10^{19}	2.1
D1	n-type	9.69×10^{19}	0.727
D2	n-type	5.6×10^{18}	31.4
D3	n-type	1.36×10^{20}	12.44
D4	n-type	4×10^{18}	54.3

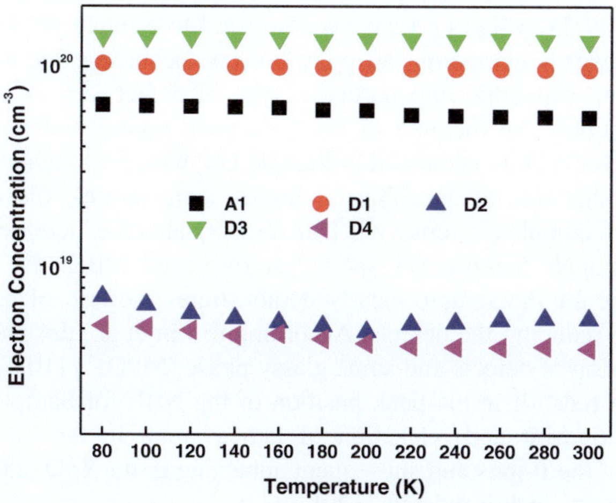

Fig. 3.18 Temperature-dependent electron carrier concentration for the various samples

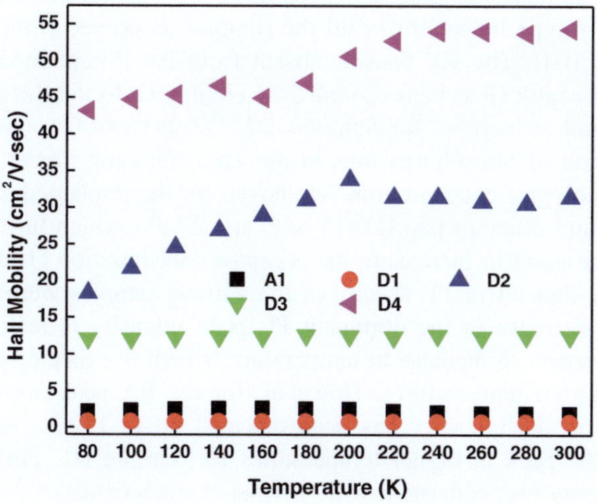

Fig. 3.19 Temperature-dependent Hall mobility of the obtained samples

Optical Properties

The optical characteristics of the samples were accomplished by low-temperature (8 K) PL study (Fig. 3.20). A prominent NBE at around 3.38 eV with no deep-level emission was noticed for Sample A1, verifying the growth of high-quality ZnO film. For Sample D1, a decrease in the NBE peak intensity is detected. In addition, a broad deep-level emission centred at 2.55 eV is observed. The large FWHM value of this peak indicates that this emission derives from the origin of different point defects occurring due to P-implantation, along with the Zn_i and V_O. An increase in the intensity of the NBE and deep-level emission is observed for Sample D2 along with a lower FWHM value (1.4 eV) as compared to Sample D1 (3.9 eV). When annealing at lower temperature, the point defects, occurring due to implantation, might have been converted into particular type of defect (Zn_i or V_O), explaining such an observation. An increase of the NBE peak intensity along with the subsidised deep-level peak is witnessed in Sample D3, which was annealed at 750 °C. This indicates that due to annealing at higher temperatures, the defects due to implantation are completely removed. This also explains the increase in intensity of the NBE. Although Sample D4 shows an improved NBE, the occurrence of deep-level peak for the sample clearly demonstrates the role of background gas environment in reducing the defects. According to Lin et al., this role is correlated with the formation of defects and some glassy phase (MPO_4^-) [10]. It is interesting to note a slight redshift in the peak position of the NBE for Samples D2, D3 and D4. The development and relaxation of lattice strain in the films, due to the incorporation of the P-ions and subsequent annealing in the ZnO lattice, might be a possible reason for such a redshift [120].

Inset of Fig. 3.20 shows the PL spectra near the NBE region for the samples. A prominent donor-bound-exciton (D°X) peak at approximately 3.36 eV is seen for the implanted samples, while the FX peak at 3.38 eV is dominant in Sample A1, confirming the n-type behaviour of all the samples as proved from the Hall measurements [72, 104]. The FX peak is absent from the P-implanted samples. The free-electron–acceptor (FA) peak around 3.31 eV is seen for all the samples, but its intensity is found to increase for Samples D1, D2, D3 and D4, clearly indicating that incorporation of phosphorus ions in the ZnO films does have some effect in increasing the acceptor concentration. Moreover, for the implanted sample, we also see a donor-bound-acceptor pair (DAP) peak at 3.27 eV, which further clarifies the role of P-implantation in increasing the acceptor concentration [104].

Temperature-dependent PL spectra of the various samples were also measured (Fig. 3.21). A decrease in the dominant PL peak intensity is realised for all the samples with respect to increase in temperature, which is expected due to increase of defects at higher temperatures. However, for the FA peak around 3.31 eV, a slight increase in the intensity has been detected up to 150 K, followed by the quenching of the peak at higher temperatures for Sample D1. For the remaining samples, the decay was consistent with thermal characteristics of the FA transition [35]. The optical binding energy due to the acceptors can be predicted from the FA transition using Eq. 3.2:

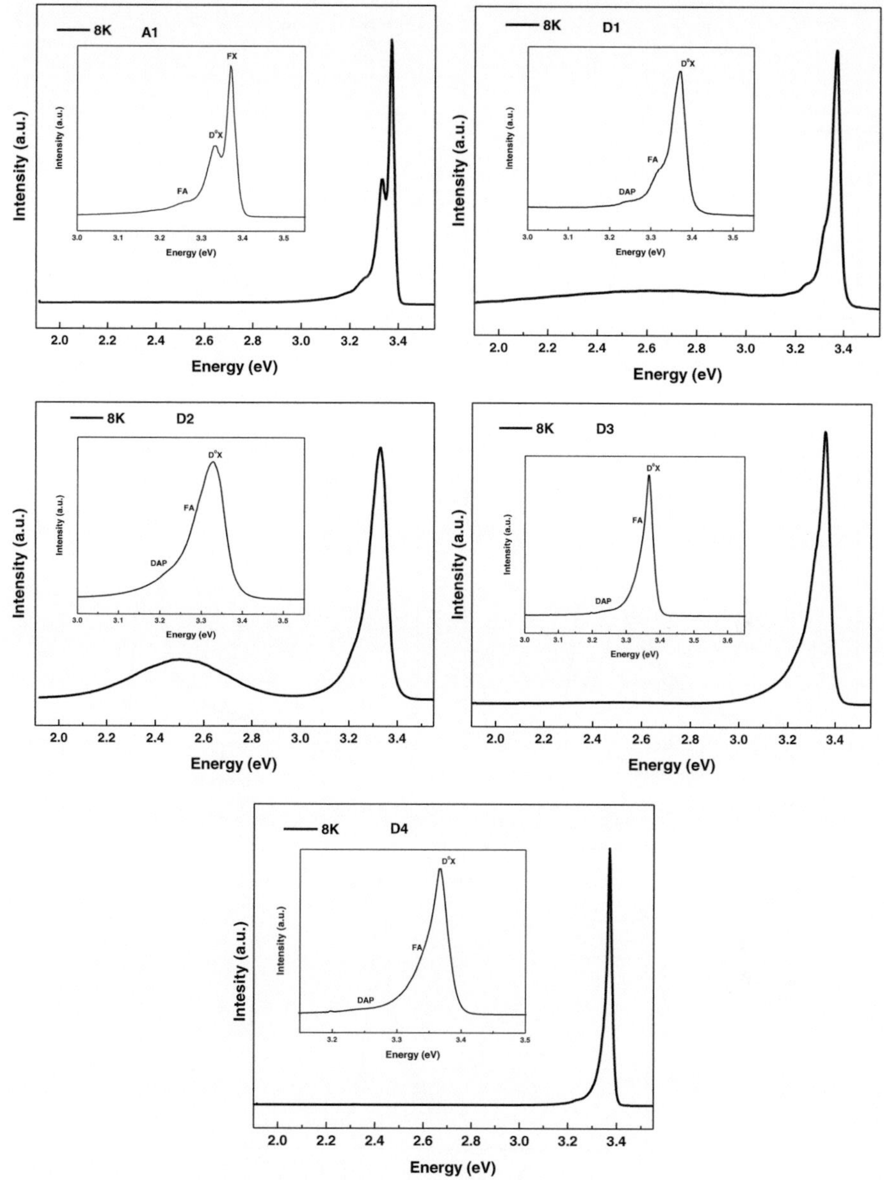

Fig. 3.20 PL spectra of different samples taken at 8 K

$$E_A = E_g - E_{FA} + k_B T/2 \tag{3.2}$$

where E_A, E_g and E_{FA} are the acceptor, bandgap and FA energies, respectively. Neglecting the thermal energy term at temperature of 8 K and using the standard

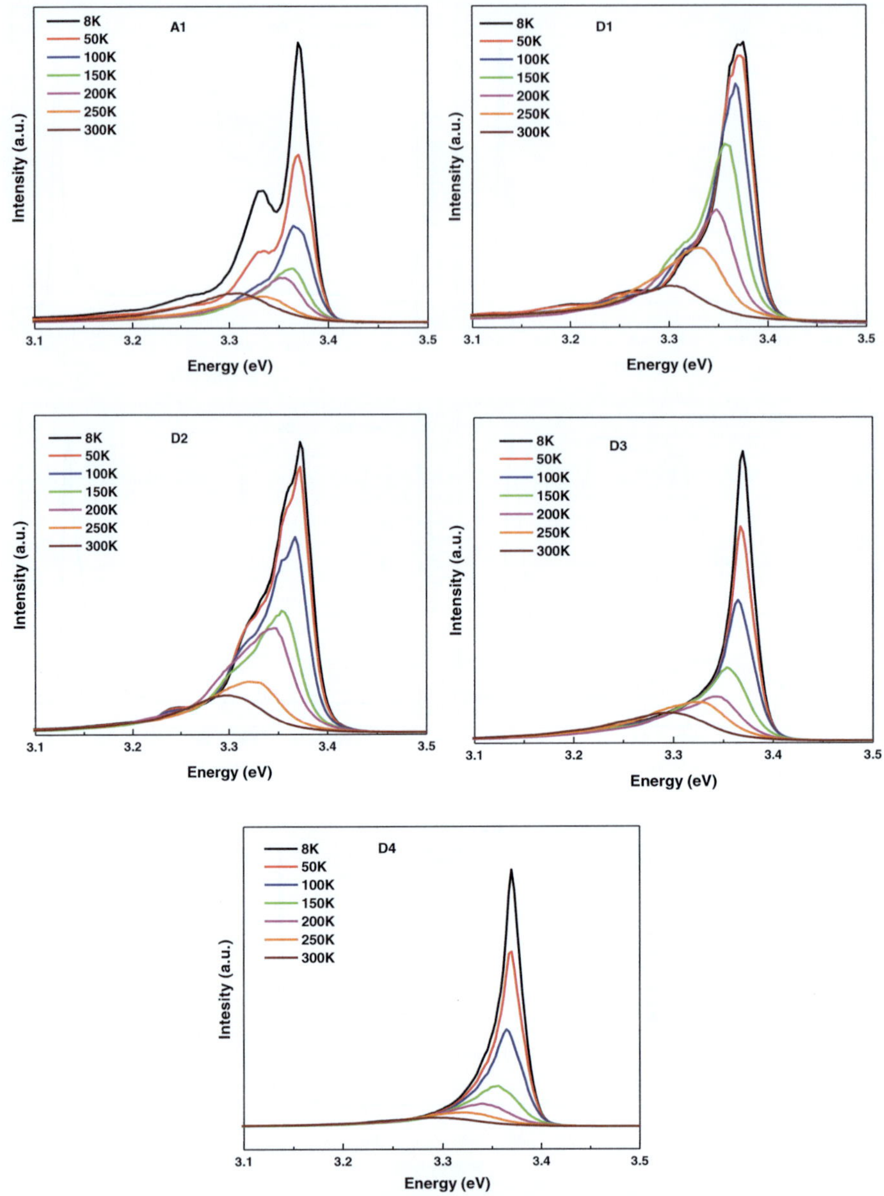

Fig. 3.21 Temperature-dependent PL spectra of different samples (Figures A1, D1, D2 and D3 reprinted from S. Nagar et al. "Effect of phosphorus irradiation in the structural, electrical and optical characteristics of ZnO thin films", *Journal of Luminescence*, Vol. 132, pp. 1089–1094, 2012; with permission from Elsevier)

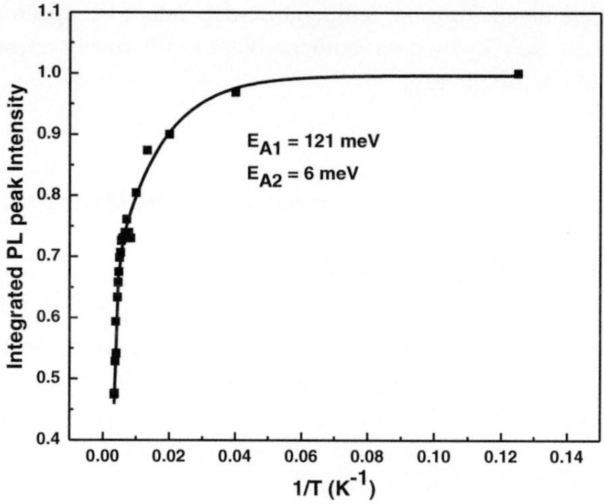

Fig. 3.22 Normalised PL peak intensity of the FA peak for Sample D3 plotted against the inverse temperature (Reprinted from S. Nagar et al. "Effect of phosphorus irradiation in the structural, electrical and optical characteristics of ZnO thin films", *Journal of Luminescence*, Vol. 132, pp. 1089–1094, 2012; with permission from Elsevier)

value of bandgap energy (E_g = 3.437 eV), the acceptor binding energy was calculated to be around 120 meV. The temperature dependence of the integrated PL peak intensity for the FA peak for Sample D3 is shown in Fig. 3.22, which was analysed using Eq. 3.3:

$$I = I_0/\{1 + a_1 \exp(-E_{a1}/k_B T) + a_2 \exp(-E_{a2}/k_B T)\} \tag{3.3}$$

where I_0 is the integrated PL peak intensity at 8 K, E_{ai} (i = 1, 2) are the activation energies of the thermal quenching processes, and a_1 and a_2 are constants. The energies of E_{a1} and E_{a2} were calculated to be 121 and 6 meV, respectively. In Eq. 3.3, E_{a1} corresponds to the activation energy of the FA peak and E_{a2} is associated with the temperature-dependent carrier capture cross-section of the recombination centres [41]. Similar curves were also obtained for other samples. Thus, phosphorus can be used as an acceptor to obtain p-type ZnO films. However, implantation parameters and the post-implantation annealing temperatures need further optimisation to obtain p-type ZnO films.

3.2.3.3 Conclusion

An attempt to make p-type ZnO by using phosphorus ion implantation was carried out at a low energy of 50 keV with fluence 8×10^{14} ions/cm^2. Highly c-axis-oriented ⟨002⟩ ZnO films were obtained. The results of PL measurements showed an increase in the defects due to implantation, which then reduced when the samples were heated at a high temperature of 750 °C. The increase in the FA and DAP peak intensities for the implanted samples indicated that the acceptor concentration in the ZnO film had increased. Therefore, the approach used in this study can be employed to achieve p-type ZnO. The activation energy of the FA peak was calculated from the temperature-dependent PL spectra and was found to be around 120 meV, which is comparable to previously reported results of the acceptor activation energy.

3.3 Plasma Immersion Ion Implantation Studies

3.3.1 Deposition of ZnO

Undoped ZnO thin films were deposited using radio frequency (RF) magnetron sputtering on highly resistive ⟨100⟩ n-type silicon substrates. A high-purity ZnO target (99.999%) with a 5 inch diameter was used. The target-to-substrate distance was kept at 70 mm. The deposition was performed for 40 min in an Ar ambient of 7×10^{-3} mbar at room temperature. A base vacuum of 2×10^{-5} mbar was achieved prior to the deposition. A resultant thickness of 350 nm, measured through ellipsometry, was achieved.

3.3.2 Phosphorus Implantation Studies

3.3.2.1 Experimental Details

The samples were subsequently subjected to phosphorus implantation using the plasma immersion ion implantation (PIII) technique to obtain Sample P1. A high-voltage negative pulse of 2 kV was applied to the target for 30 s. The negative pulse had an ON time of 10 μs and a RF of 5 kHz. The chamber pressure of 1×10^{-1} mbar was achieved from phosphine (PH_3) gas during implantation, which was carried out using RF plasma power of 900 W. After implantation, the samples were rapid thermal annealed at 700, 800, 900 and 1000 °C in oxygen environment for 10 s to eradicate implantation-related defects from the sample to obtain Samples P2, P3, P4 and P5 respectively.

3.3.2.2 Results and Discussion

Structural Properties

Figure 3.23 shows the SEM images of the different samples. The presence of defects due to phosphorus implantation in Sample A can be clearly observed. With an increase in annealing temperature, grains grow, and large grains in the films are clearly visible for Sample E, implying the removal of any implantation-related defects.

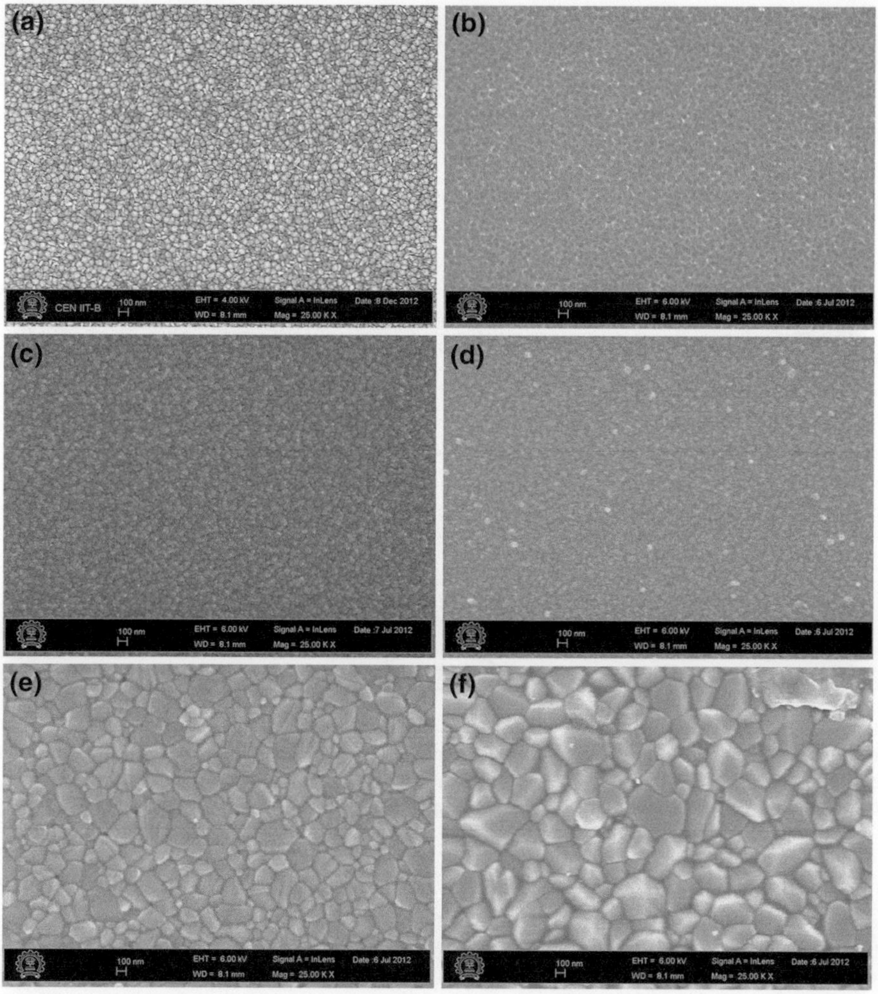

Fig. 3.23 SEM images of the obtained samples **a** as-deposited, **b** Sample P1, **c** Sample P2, **d** Sample P3, **e** Sample P4 and **f** Sample P5 (Figures (**b**–**e**) reprinted from S. Nagar et al. "Evidence of strong acceptor peaks in ZnO thin films doped with phosphorus by plasma immersion ion implantation technique", *Journal of Luminescence*, Vol. 137, pp. 55–58, 2013; with permission from Elsevier)

Optical Properties

Low-temperature PL (8 K) spectra of various samples are shown in Fig. 3.24. An expected donor-bound-exciton (D°X) peak around 3.36 eV is observed for Sample P1 along with deep-level defect peaks centred at 2.5 eV corresponding to the Zn_i, V_O and implantation-related defects [72, 121]. For Sample P2, the NBE almost

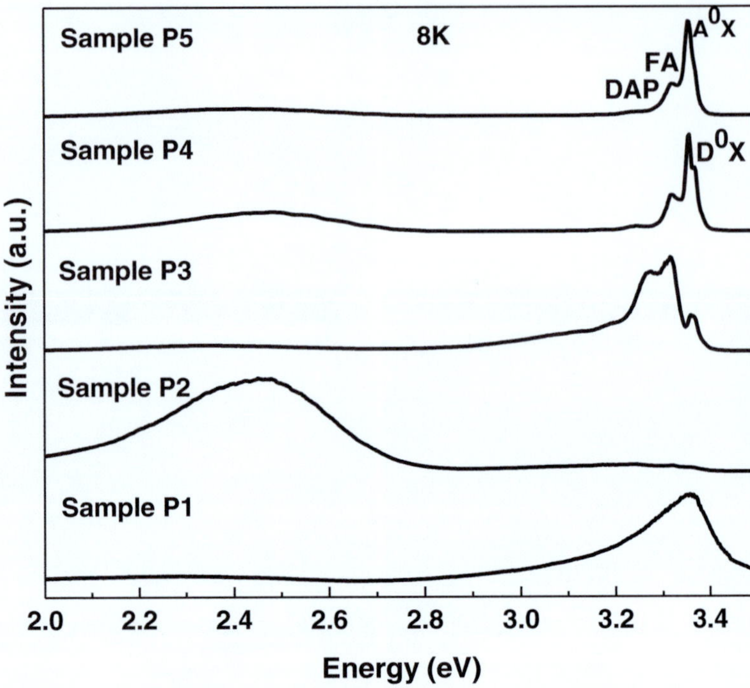

Fig. 3.24 PL spectra of different samples measured at 8 K showing the evolution of A°X peak with increase in annealing temperature (Reprinted from S. Nagar et al. "Evidence of strong acceptor peaks in ZnO thin films doped with phosphorus by plasma immersion ion implantation technique", *Journal of Luminescence*, Vol. 137, pp. 55–58, 2013; with permission from Elsevier)

disappears, while the deep-level defect peak becomes dominant, signifying the increased number of defects in the sample. Upon annealing temperature increase, the NBE again starts to dominate the PL spectra as seen in Samples P3, P4 and P5. However, a noteworthy change in the emissions is visible for the different samples. Whereas a FA peak around 3.32 eV is dominant in Sample P3, the A°X peak around 3.35 eV becomes dominant for Samples P4 and P5. This result clearly depicts a change in carrier type from n-type for Samples P1 and P2 to p-type for Samples P3, P4 and P5 due to P-implantation [49, 72, 121]. Since the implanted samples are annealed at higher temperatures, the P-ions receive sufficient thermal energy to occupy the V_O positions in the ZnO lattice and form shallow acceptors, thus increasing the acceptor concentration and giving rise to p-type films. To further validate the importance of phosphorus implantation and subsequent annealing, PL measurements were performed on the as-deposited and annealed samples without implantation (Fig. 3.25). The absence of any acceptor peaks in these samples proves that implantation of P-ions leads to an increase in acceptors and thus p-type

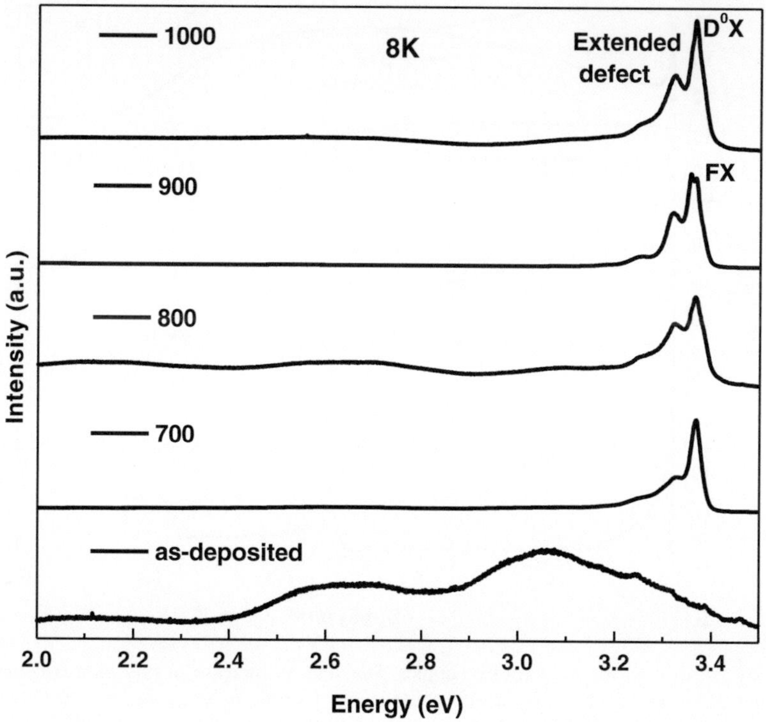

Fig. 3.25 PL spectra of unimplanted and annealed samples measured at 8 K. No A°X peak is visible in the samples (Reprinted from S. Nagar et al. "Evidence of strong acceptor peaks in ZnO thin films doped with phosphorus by plasma immersion ion implantation technique", *Journal of Luminescence*, Vol. 137, pp. 55–58, 2013; with permission from Elsevier)

film formation. The presence of the FA peak around 3.32 eV in Samples P4 and P5 and the DAP peak around 3.25 eV in Samples P3, P4 and P5 further confirms the increase in acceptors. Moreover, with regard to annealing the samples at higher temperatures, the defect peaks are also minimised, revealing reduced numbers of acceptor-, donor-, and implantation-related defects. This clearly illustrates that annealing the samples at higher temperatures improves film quality and changes carrier type.

To clearly visualise the effect of annealing on the P-ions, a secondary ion mass spectroscopy (SIMS) study was performed on the samples (Fig. 3.26). Increased P counts along the depth of the ZnO film can be clearly observed at higher temperatures. This finding further supports the PL and SEM results discussed above. In Sample P5, it is observed that some probable intermixing between the ZnO and Si occurs. Since the n-type Si was doped with phosphorus atoms, it may be possible that the P atoms from Si diffuse into the ZnO films and increase its count. However,

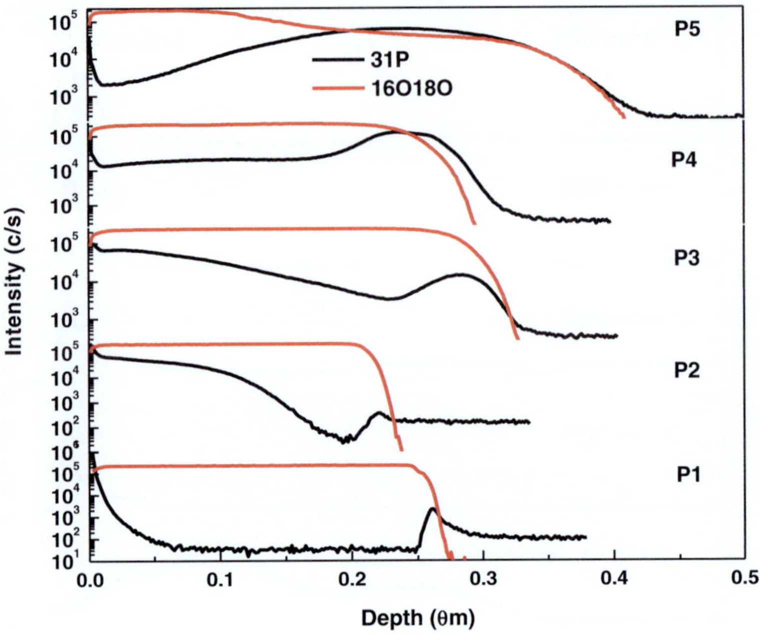

Fig. 3.26 SIMS data for the various samples depicting an increase in phosphorus counts with annealing (Reprinted from S. Nagar et al. "Evidence of strong acceptor peaks in ZnO thin films doped with phosphorus by plasma immersion ion implantation technique", *Journal of Luminescence*, Vol. 137, pp. 55–58, 2013; with permission from Elsevier)

it is not contributing towards p-doping, i.e. acceptor concentration, because PL of unimplanted annealed samples did not exhibit any acceptor peak. To further confirm the p-type behaviour of the samples, Van der Pauw Hall measurements were attempted, but the high resistivity of the samples made such measurements impossible.

Temperature-dependent PL measurements for Sample P5 are shown in Fig. 3.27. While the dominant A°X peak diminishes exponentially with temperature, the FX peak dominates at higher temperatures. The increasing number of FXs at higher temperatures leads to the following observation. The exponential decay of the acceptor peak with temperature is shown in the inset of Fig. 3.27. Quantitative fitting of this peak was performed using Eq. 3.4:

$$I = I_0/[1 + A * \exp(-E_A/k_B T)] \quad (3.4)$$

where I_0 is the integrated PL peak intensity at $T = 8$ K, E_A is the acceptor activation energy, and A is the peak fitting constant. The acceptor activation energy was thus determined to be 125 meV. This finding was similar to that obtained using spectroscopic observation where Eq. 3.2 was applied. The acceptor activation energy

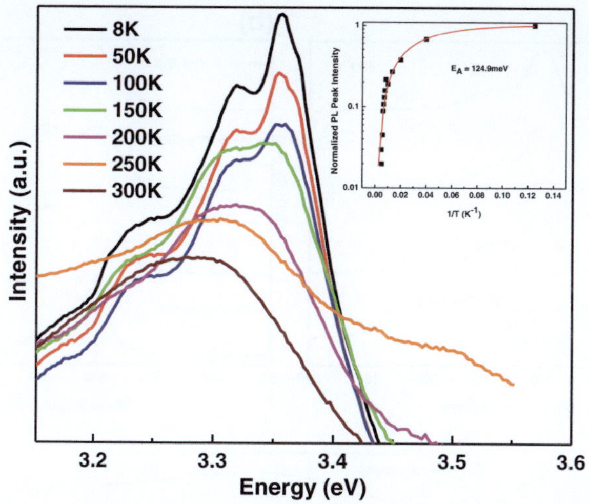

Fig. 3.27 Temperature-dependent PL spectra of Sample P5. *Inset* variation of acceptor peak intensity with temperature for Sample P5. The acceptor activation energy was calculated to be 125 meV (Reprinted from S. Nagar et al. "Evidence of strong acceptor peaks in ZnO thin films doped with phosphorus by plasma immersion ion implantation technique", *Journal of Luminescence*, Vol. 137, pp. 55–58, 2013; with permission from Elsevier)

was measured to be 120 meV, which is similar to the value calculated above. The shallow acceptor activation energy proves that P acts as a good dopant for producing p-type ZnO thin films.

Reliability Studies

Low-temperature PL (8 K) has been performed on the p-type samples over a period of 8 months to check the stability of the attained p-type films with time (Fig. 3.28). In the case of Sample P5, a dominant FA peak is observed after 4 and 6 months, while for Sample P4, the FA peak is visible even after a period of 8 months. However, for Sample P5, another peak corresponding to A°X peak is observed after 4 months, while a D°X peak is seen after 6 months. In P4, another peak corresponding to D°X peak is visible even after 4 months, and no A°X could be observed. This finding clearly indicates that the Sample P5 was still p-type after 4 months, but after 6 months, it starts converting back to n-type due to self-compensation, while Sample P4 starts to convert to n-type even after 4 months. However, after a period of 8 months, Sample P5 has deteriorated and thus no significant PL emission is observed. In the case of Sample P3, a strong D°X peak is observed even after a period of 4 months, which indicates that the ZnO film has completely converted to n-type after the given period.

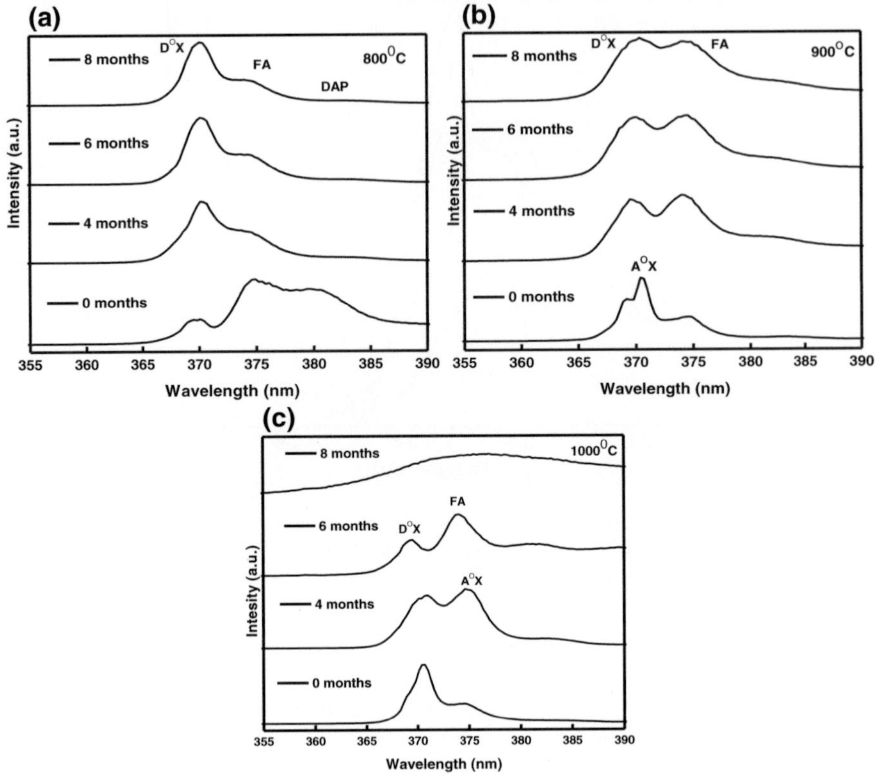

Fig. 3.28 PL spectra of **a** P3, **b** P4 and **c** P5, taken over a period of 8 months

3.3.2.3 Conclusion

PIII technique was used to implant P-ions on 350-nm ZnO thin films to create p-type ZnO films. The implanted samples are subsequently annealed at 700–1000 °C. Reduced numbers of defects and proper grain formation with increasing annealing temperature were visible in the SEM images. The incorporation of P-ions in the film was visualised in the SIMS study. The low-temperature PL studies revealed a gradual transition from n-type to p-type characteristics, with increasing annealing temperature. A strong acceptor peak for the sample annealed at 1000 °C with shallow acceptor activation energy of 125 meV confirmed the formation of p-type ZnO. The p-type behaviour of the thin film was maintained up to a period of 6 months, although the intensity reduced drastically. However, with better optimisation of deposition and implantation parameters, it may be possible to increase the reliability of the obtained p-type films. These results can help with the successful fabrication of future ZnO-based optoelectronic devices, such as LEDs and LDs.

3.3.3 *Nitrogen Implantation Studies*

3.3.3.1 Experimental Details

Nitrogen ion implantation was carried out on the ZnO films, deposited by RF sputtering, using PIII to obtain Sample N1. An RF of 5 kHz was applied on a high-voltage negative pulse of 2 kV with a pulse width of 10 μs. The ion implantation was carried out for 30 s in nitrogen plasma under a power of 900 W. A 2×10^{-5} mbar vacuum was achieved in a high-vacuum chamber prior to implantation. During implantation, the vacuum pressure was increased to 1×10^{-1} mbar. It is known that ion implantation induces defects in films. To reduce or remove such defects, the ion-implanted film samples were subjected to rapid thermal annealing in an oxygen atmosphere at 700, 800, 900 or 1000 °C for 10 s. As a result, Samples N2, N3, N4 and N5, respectively, were prepared.

3.3.3.2 Results and Discussion

Structural Properties

SEM images of Samples N1–N5 are shown in Fig. 3.29. A high number of defects were observed in Samples N1 and N2. Grain formation occurs upon ion implantation and annealing, and grain size increased with annealing temperature. Large grains are visible in Sample N5. These results indicate the removal of implantation-induced defects induced via annealing.

Optical Properties

Low-temperature (8 K) PL spectra for the samples are shown in Fig. 3.30. While a NBE is visible for as-deposited unimplanted sample, it is not visible in Samples N1 and N2. However, deep-level defect peaks around 1.9 and 2.5 eV, corresponding to V_O, Zn_i and implant-related defects, can be seen in the spectra of these samples [72, 121]. The occurrence of defect peaks in Samples N1 and N2 supports the observation of defects in the SEM images (Fig. 3.29a, b). In annealed Samples N3–N5, NBEs are clearly observed along with reduced deep-level defect peaks, which support the reduction of defects and increase in grain formation observed in these samples (Fig. 3.29c–e). However, there were significant changes in the PL spectra of some samples. A dominant FA peak at ~3.32 eV is present in Sample N3 along with peaks at 3.25–3.36 eV, which correspond to the DAP peak and D°X peak, respectively. This suggests that nitrogen implantation leads to an increase in acceptors in ZnO film. Upon further increasing the annealing temperature (i.e. Samples N4 and N5), A°X peaks, around 3.35 eV, dominate the spectra. In the PL spectra for the as-deposited and annealed samples without any implantation, no

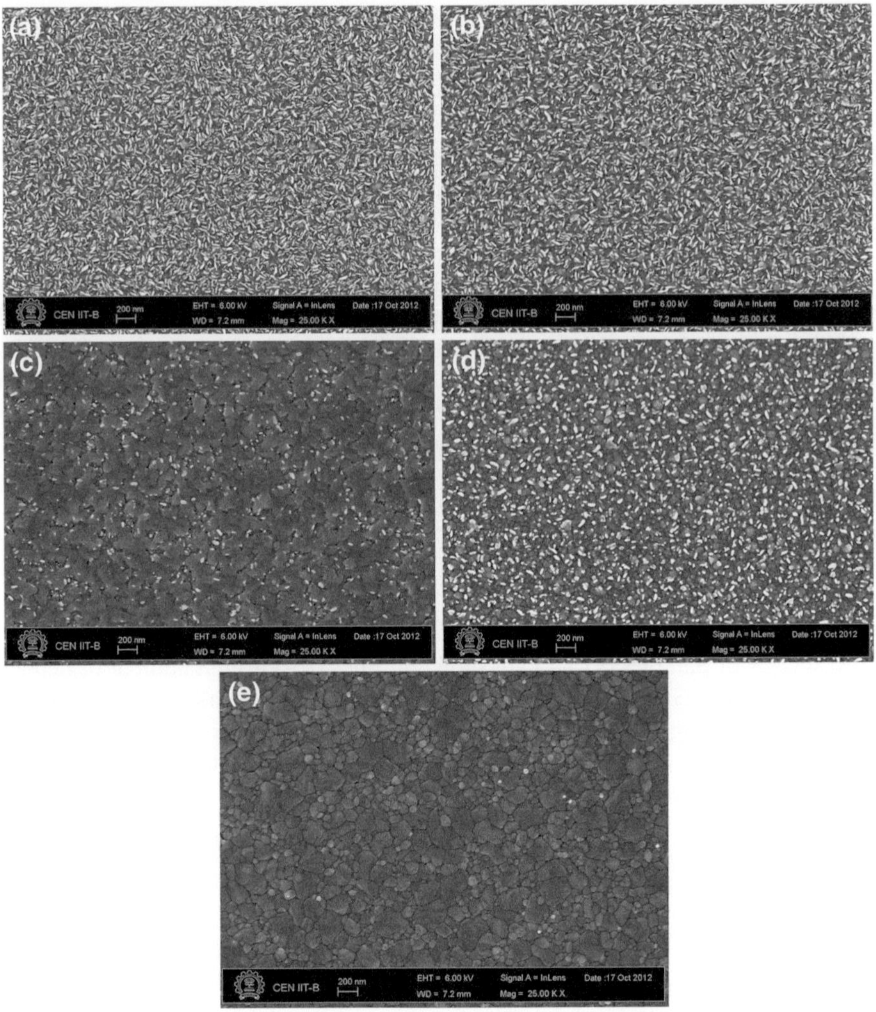

Fig. 3.29 SEM images of the different samples: **a** Sample N1, **b** Sample N2, **c** Sample N3, **d** Sample N4 and **e** Sample N5 (Reprinted from S. Nagar et al. "Realization of reliable p-type ZnO thin films by nitrogen implantation using plasma immersion ion implantation", *Superlattices and Microstructures*, Vol. 75, pp. 9–16, 2014; with permission from Elsevier)

dominant acceptor peaks were visible (Fig. 3.25). These spectra show the importance of nitrogen implantation in achieving p-type ZnO films. In addition, Fig. 3.30 shows that FA and DAP peaks are also visible in the implanted and annealed samples. While a D°X peak appears as a shoulder peak in Sample N4, no such peak was observed in Sample N5. Thus, we infer that Samples N4 and N5 have converted into p-type. This conversion takes place because at higher temperatures, the

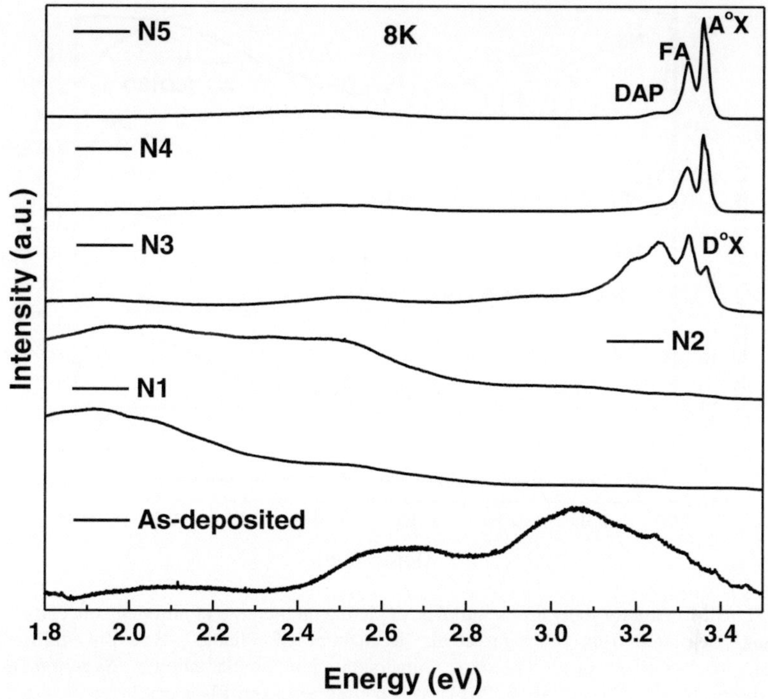

Fig. 3.30 PL spectra of Samples N1–N5 taken at 8 K. Marked A°X peaks can be observed at ~3.35 eV in the spectra for Samples N4 and N5

nitrogen atoms in the film are activated and occupy lattice sites, thus contributing to an increase in acceptors in the films and subsequently forming a p-type film.

A SIMS assessment was undertaken to determine the effect of annealing on the implanted nitrogen ions and to determine the nitrogen concentration in each of the samples (Fig. 3.31). Sample nitrogen concentration was determined after calibration by using a standard N/ZnO sample to acquire quantitative information. The results showed an increase in nitrogen concentration from 1×10^{18} cm^{-3} in Sample N1 to 3×10^{18} cm^{-3} in Sample N5 near the surface. Moreover, at higher annealing temperatures, nitrogen diffusion to greater depths is evident. Although SIMS cannot specify the lattice sites that are occupied by the implanted atoms, the PL results hint that nitrogen atoms in the film are activated and occupy lattice sites, thus contributing to an increase in acceptors in the films and subsequently forming a p-type film for higher annealing temperatures.

The PL spectra recorded at various temperatures between 8 K and 300 K are shown in Fig. 3.32. The dominant A°X peak detected at low temperatures decreases exponentially with increasing temperature to give rise to an FX peak at higher temperatures. The presence of large number of free excitons at high temperatures leads to the dominance of this peak at those high temperatures. However, the FA peak remains visible throughout the temperature range tested. Figure 3.32 inset

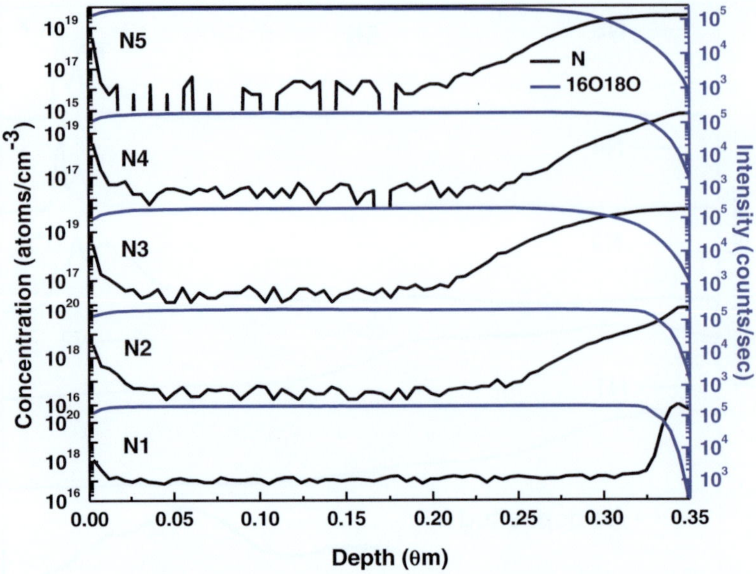

Fig. 3.31 SIMS data for the various samples, depicting an increase in nitrogen concentration with annealing (Reprinted from S. Nagar et al. "Realization of reliable p-type ZnO thin films by nitrogen implantation using plasma immersion ion implantation", *Superlattices and Microstructures*, Vol. 75, pp. 9–16, 2014; with permission from Elsevier)

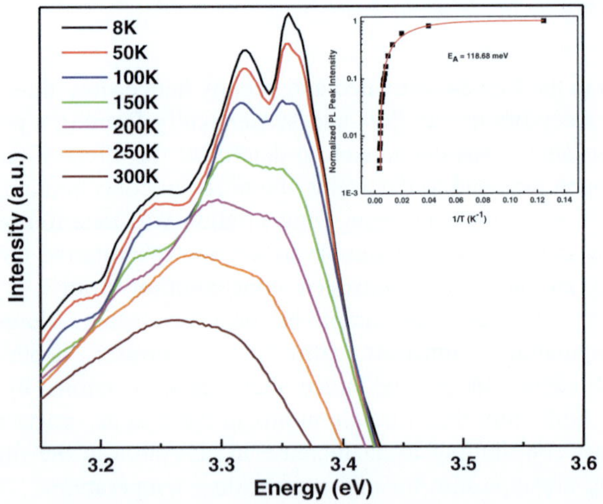

Fig. 3.32 Temperature-dependent PL spectra for Sample P5. *Inset* shows a decrease in acceptor peak intensity with increasing temperature. Acceptor activation energy was 118 meV (Reprinted from S. Nagar et al. "Realization of reliable p-type ZnO thin films by nitrogen implantation using plasma immersion ion implantation", *Superlattices and Microstructures*, Vol. 75, pp. 9–16, 2014; with permission from Elsevier)

shows the normalised intensity of the A°X peak with respect to the inverse temperature. The curve was fitted using Eq. 3.4. The acceptor activation energy (E_A) was thus calculated to be around 118 meV. This shallow acceptor activation energy further indicates that the annealed implanted samples are p-type in nature.

Reliability Studies

The stability of the obtained p-type ZnO thin films was checked over a period of 9 months by performing low-temperature PL (8 K) measurements (Fig. 3.33). There was not much change in the PL spectra of any of the p-type samples. Samples N4 and N5 continue to show a dominant A°X peak along with the FA peak, while the dominant peak in Sample N3 is the FA peak. These are one of the few results

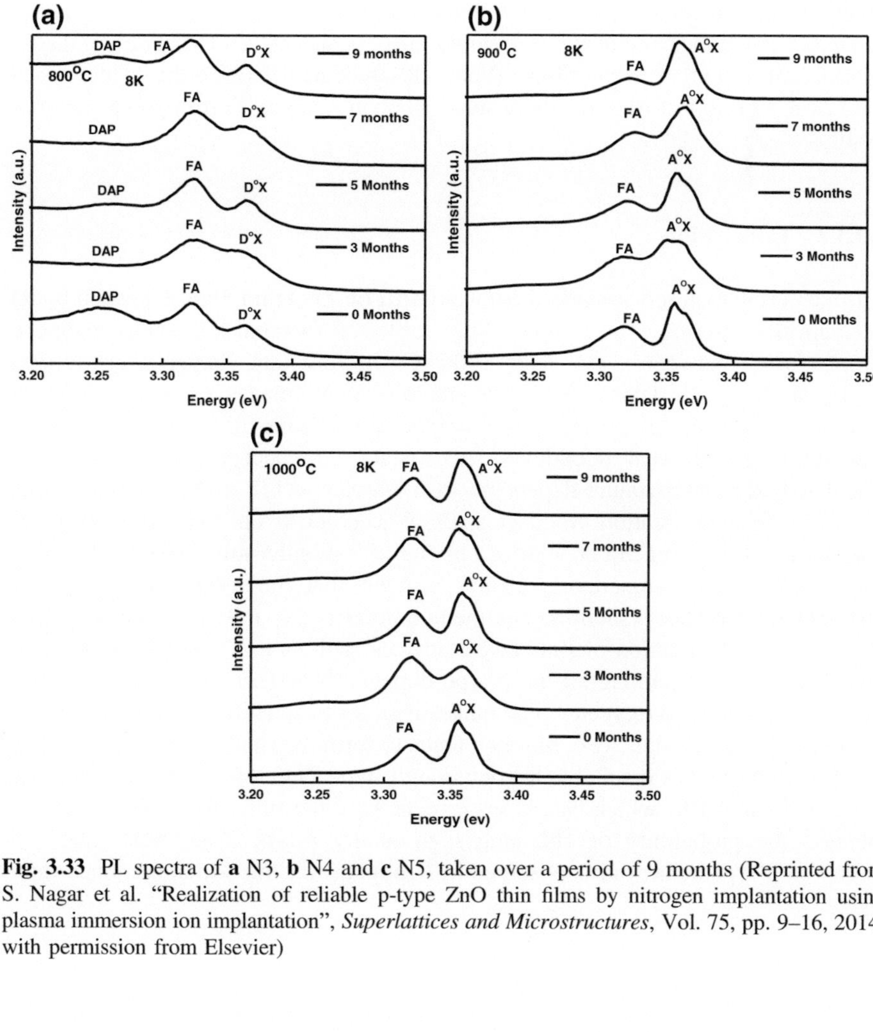

Fig. 3.33 PL spectra of **a** N3, **b** N4 and **c** N5, taken over a period of 9 months (Reprinted from S. Nagar et al. "Realization of reliable p-type ZnO thin films by nitrogen implantation using plasma immersion ion implantation", *Superlattices and Microstructures*, Vol. 75, pp. 9–16, 2014; with permission from Elsevier)

Table 3.8 A comparison of the reliable p-type ZnO obtained by the authors with the published reports

Authors	p-type dopant	Characterisation technique	p-type stability duration
Nagar et al.	P	PL	4–6 months
	N	PL	9 months
Barnes et al. [122]	N	Hall	1–2 days
Lu et al. [123]	Li–N	Hall	3–4 months
Kim et al. [42]	P	Hall, PL	8 months
Lin et al. [19]	Na	Hall, PL	11 months
Allenic et al. [124]	P	Hall	18 months

which demonstrate the reliability of p-type ZnO over a period of time. However, most of the reliability results are demonstrated using Hall measurement which is not a very reliable technique. A much more reliable technique to demonstrate p-type behaviour is PL measurements wherein the peak positions can easily determine the presence of acceptors in the films. Table 3.8 tabulates the reliability of the p-type ZnO over a period of time for the results obtained along with some other reported results.

3.3.3.3 Conclusion

Nitrogen implantation was carried out using PIII on ZnO thin films deposited by RF magnetron sputtering. The samples were subjected to rapid thermal annealing at temperatures between 700 and 1000 °C in ambient oxygen to remove implantation-related defects that were visible in SEM images. PL spectra taken at 8 K clearly showed the presence of strong acceptor peaks for samples annealed at 900–1000 °C with weak deep-level defect peaks. The acceptor activation energy calculated from temperature-dependent PL data was $\sim$118 meV, which confirms that nitrogen incorporation produces shallow acceptor levels and results in p-type ZnO films. The PL spectra done after a period of 9 months still showed a dominant A°X peak at low temperature, clearly proving that the p-type behaviour was maintained even after 9 months. The method described in this study would help in the successful fabrication of ZnO-based devices such as LEDs and LDs. As mentioned in Sect. 1.2, among all the p-type dopants, N is best suited because of its similar atomic size to oxygen which makes it easier to substitute without leading to any defect in the crystal. Also, nitrogen tends to form the shallowest acceptor levels in ZnO which makes it suitable for obtaining p-type ZnO. Moreover, nitrogen was implanted using PIII which causes lesser defects in the films. These lesser defects increase the probability of obtaining high-quality p-type ZnO films. Thus, an improved p-type behaviour is observed in N-implanted films which are much more stable.

Chapter 4
Fabrication of Optoelectronics Devices

Abstract Once reliable and reproducible p-type ZnO was achieved, different heterojunction and homojunction devices were fabricated. The heterojunction p–n junction diode was fabricated by depositing p-type ZnO films over low resistivity n-Si substrates. The p-type ZnO was achieved by phosphorus as well as nitrogen implantation. The current–voltage characteristics of the fabricated p–n heterojunction diode showed a clear rectifying behaviour with a threshold voltage of 3.3 V was observed for the phosphorus-implanted device. For the nitrogen-implanted device, a threshold voltage of 1.7 V was attained. This low threshold voltage may help to increase the lifetime of the devices. Homojunction LED was fabricated over high resistivity n-Si substrates, such that the effect of substrate conduction is nullified. The p-type ZnO was deposited by the phosphorus and nitrogen implantation as described above. The n-type ZnO was deposited over the p-ZnO by PLD technique using the optimised temperature and oxygen parameters of 650 °C and 40 mTorr, respectively. Room temperature electroluminescence (RTEL) of the fabricated device (phosphorus implantation) clearly depicted the emission of UV light around 3.18 eV. However, EL emission is dominated by a broad emission around 1.8 eV. This observation is based on the presence of deep-level defects in ZnO films. RTEL for the device fabricated by nitrogen-implanted p-ZnO showed a broad emission around 1.8 eV, corresponding to the red emission, is observed from the device. However, no UV emission was observed from the fabricated device.

Keywords Device fabrication · Heterojunction diode · Homojunction diode · Electroluminescence

Part of the chapter has been reprinted from[S. Nagar et al. "Realization of reliable p-type ZnO thin films by nitrogen implantation using plasma immersion ion implantation", Superlattices and Microstructures, Vol. 75, Pg 9-16, 2014]

S. Nagar and S. Chakrabarti, *Optimisation of ZnO Thin Films*,
DOI 10.1007/978-981-10-0809-2_4

4.1 Introduction

As mentioned in Chap. 1, the main aim of our work is to fabricate ZnO-based optoelectronics devices once the high quality and reliable p-type ZnO are achieved. In this chapter, we look into the fabrication of these devices. Fabrication of a semiconductor device requires many essential processes such as device fabrication, mask preparation, lithography, etching, lift-off and metallisation, among others.

4.2 Fabrication Steps

Flow chart of the fabrication process is shown in Fig. 4.1.

- **Cleaning of substrate**: The silicon wafer was first subjected to the common RCA cleaning for removal of organic contamination, thin oxide layer and ionic contamination.
- **ZnO film growth**: The p-type ZnO was deposited over the RCA-cleaned silicon substrate by the processes, described in Sects. 3.3.2 and 3.3.3. The n-type ZnO thin films were deposited over p-ZnO by PLD using the optimised values of temperature and oxygen pressure described in Sects. 2.2 and 2.3.
- **Cleaning of devices**: The device thus prepared was degreased using acetone, IPA and DI water for 3 min each and dried using a nitrogen gas jet.
- **Mesa formation**: Optical lithography technique was used to form mesa patterns with the help of the mesa mask. The lithography parameters used for the purpose are shown in Table 4.1.

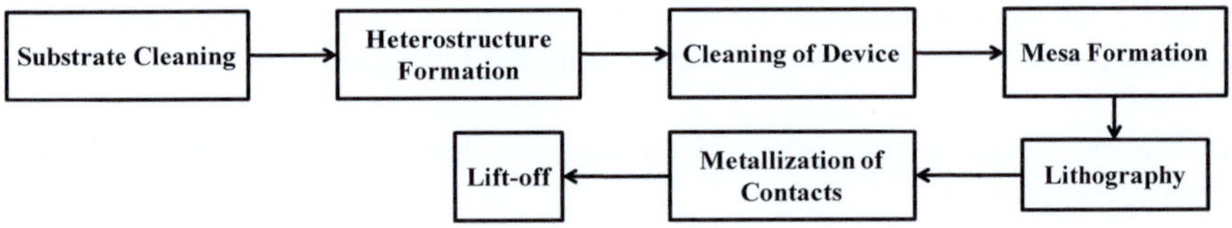

Fig. 4.1 Flow chart of processes performed during fabrication of ZnO-based devices

Table 4.1 Lithography parameters for mesa formation

Serial No.	Process	Optimised parameter
1	Photoresist	S1813
2	Spinning	3000 rpm for 30 s
3	Prebaking	90 °C for 3 min
4	Exposure	4.5 s
5	Developer	MF-319
6	Developed time	45 s

Mesa defined by lithography was then etched away using a dilute HCl (HCl: H_2O::1:250). The samples were kept in the solution for 25 s to etch away 350 nm of ZnO thin film.

- **Contact formation**: Once the mesa formation takes place, it is required to make ohmic contacts for the top and bottom contact for the proper functioning of the p–n junction diode. Once the metallisation was complete, lift-off process was carried out, for which the sample was kept in acetone for 5 min, followed by sonication.

4.3 Heterojunction Devices

4.3.1 p-ZnO/n-Si

4.3.1.1 Phosphorus-Doped p-Type ZnO

The p-type ZnO was prepared by the technique described in Sect. 3.3.2. A sample similar to Sample P5 was prepared over low-resistive n-type Si substrate and fabricated as described in Sect. 4.2. A Ni (20 nm)/Au (200 nm) was used to make contact for the p-type ZnO film, while Al was used as the contact for n-type Si [87]. Current–voltage characteristics of the fabricated p–n heterojunction diode are shown in Fig. 4.2. A clear rectifying behaviour with a threshold voltage of 3.3 V was observed.

4.3.1.2 Nitrogen-Doped p-Type ZnO

The p-type ZnO was prepared by the technique described in Sect. 3.3.3. A sample similar to Sample N5 was prepared over low-resistive n-type Si substrate and fabricated as described in Sect. 4.2. The contact used was similar to the one described in Sect. 4.3.1.1. Current–voltage characteristics of the fabricated p–n heterojunction diode are shown in Fig. 4.3. The threshold voltage of the junction was determined to

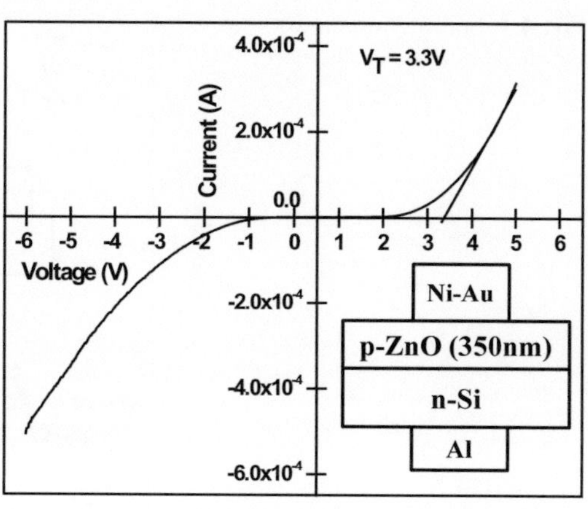

Fig. 4.2 Current–voltage (*I–V*) characteristics of a p-ZnO/n-Si heterojunction diode with a threshold voltage of 1.7 V. *Inset* shows a schematic of the fabricated heterojunction device

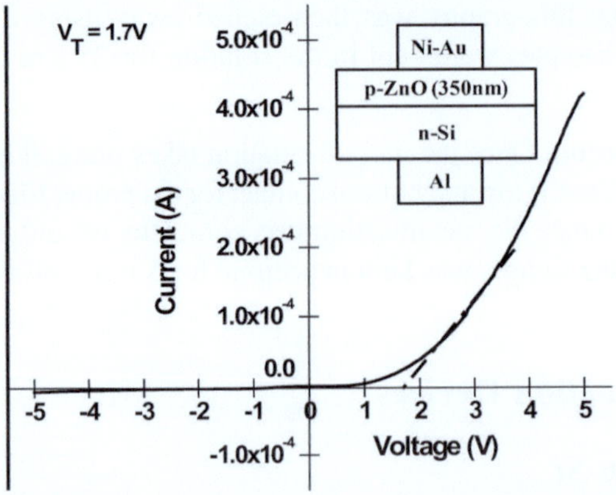

Fig. 4.3 Current–voltage (*I–V*) characteristics of a p-ZnO/n-Si heterojunction diode with a threshold voltage of 1.7 V. *Inset* shows a schematic of the fabricated heterojunction device (Reprinted from S. Nagar et al. "Realization of reliable p-type ZnO thin films by nitrogen implantation using plasma immersion ion implantation", *Superlattices and Microstructures*, Vol. 75, pp. 9–16, 2014; with permission from Elsevier)

be ~1.7 V, which is similar to previously obtained values [125, 126]. Such a low threshold voltage would increase the lifetime of such ZnO-based devices.

4.3.2 p-ZnO/p-Si Detector

The fabrication of the device was carried out as mentioned in Sect. 4.2. The p-type ZnO film was prepared through nitrogen implantation, as explained in Sect. 3.3.3. Al was used as the bottom contact for p-Si, while a Ni/Au stack was used as the contact for the p-ZnO layer. Figure 4.4 shows the spectral response of the detector thus fabricated. White light was used as the excitation source for measuring the response. Peaks at 1.03 μm, 515 nm and 344 nm were observed.

Fig. 4.4 Spectral response for the fabricated device. *Inset* the fabricated device

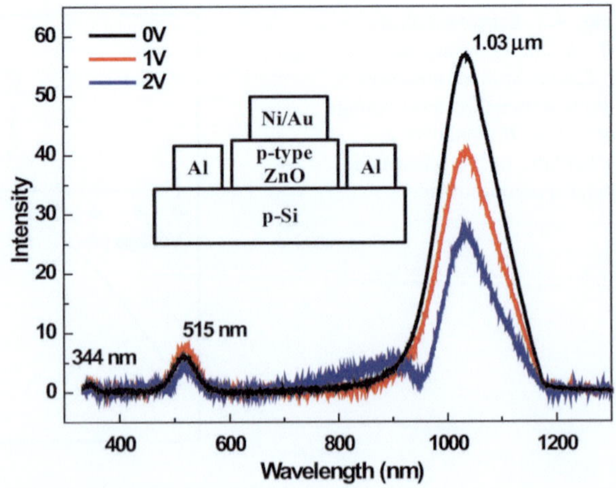

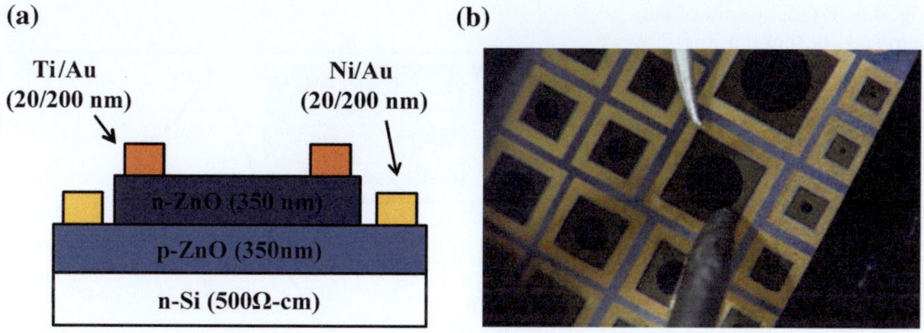

Fig. 4.5 **a** Cross section; **b** top-view of the fabricated device along with probing. The p-type ZnO is made by both P-doping as well as N-doping

4.4 Homojunction Devices

4.4.1 p-ZnO/n-ZnO Homojunction LED

The p-type ZnO was prepared using the technique described in Sects. 3.3.2 and 3.3.3. The n-type ZnO layer was deposited using the PLD technique with the optimised parameters, described in Sects. 2.2 and 2.3. The device was fabricated as described in Sect. 4.2. A Ni/Au (20/200 nm) stack was used for the p-type contacts, while a stack of Ti/Au (20/200 nm) was used as the n-type contact [87]. The cross section and top-view of the device thus fabricated is shown in Fig. 4.5.

4.4.1.1 Phosphorus-Doped p-Type ZnO

The room temperature electroluminescence (RTEL) spectrum of the fabricated device is as shown in Fig. 4.6. The snapshot of the RTEL emission is as depicted in the inset of Fig. 4.6. The RTEL spectrum was obtained by applying a forward bias current of 50 mA. Although the spectrum was dominated by the peak around 1.8 eV, corresponding to deep-level defects, a UV emission around 3.18 eV was also observed. While no emission was observed in the reverse bias condition, confirming the emission from the ZnO p–n homojunction, the red emission in the forward bias was so obvious that it was visible even with the naked eye in darkness. The dominance of the deep-level defect peak around 1.8 eV is attributed to the fact that most of the UV emission was absorbed by the thick n-type layer and also because the device structure was not optimised [19, 77, 127, 128].

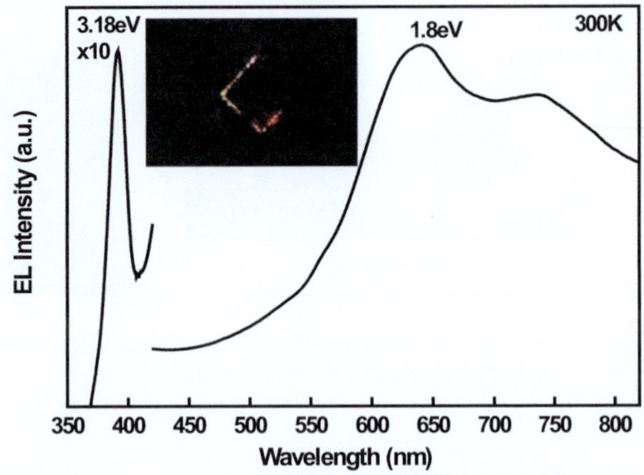

Fig. 4.6 RTEL spectra of the fabricated p-ZnO:P/n-ZnO homojunction diode. *Inset* shows the image of the emitted light from the device

4.4.1.2 Nitrogen-Doped p-Type ZnO

The RTEL spectrum shows a broad emission around 1.8 eV (Fig. 4.7). However, no UV emission was observed from the EL spectra. A forward bias current of 50 mA was applied to the device for taking the EL spectra. The fact that no EL emission was observed from the n-Si/p-ZnO heterojunction diode proves that the emission is from the p–n homojunction device. The emission around 1.8 eV may be attributed to the deep-level defects in the ZnO, to the absorption of the NBE by the thick n-layer and to the fact that the device structure was not optimised [19, 77, 127, 128].

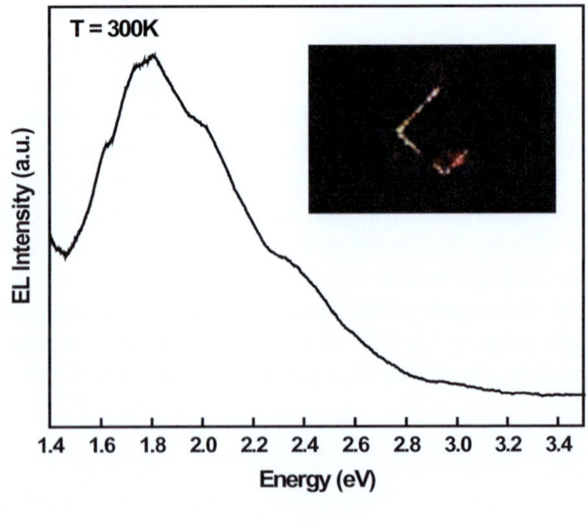

Fig. 4.7 RTEL spectra of the fabricated p-ZnO:N/n-ZnO homojunction diode. *Inset* shows the image of emitted light from the device

4.5 Conclusion

Successful fabrication of p-ZnO/n-Si heterojunction diode has been demonstrated having clear rectification behaviour. A p-ZnO/p-Si heterojunction detector with spectral response peaks at 1.03 μm, 515 nm and 344 nm has been realised. Finally, p-ZnO/n-ZnO homojunction LED has been fabricated. RTEL demonstrated dominant peak around 1.8 eV for both the homojunction devices. However, a small peak around 3.18 eV was also observed for the P-doped device. The devices fabricated and characterised above are not among the best reported in the literature [48, 76, 77, 88, 96, 97, 99, 100]. Nevertheless, attainment of highly stable p-type ZnO is one of the major steps in realising successful optoelectronics devices, and with proper optimisation of the device structure and fabrication steps, it will be possible to obtain desired results which will be at par with reported data.

Chapter 5
Summary and Future Works

Abstract The summary of the work and future work has been discussed in this chapter.

Keywords ZnMgO · Quantum wells · High-electron-mobility transistors

The wide application of ZnO in the field of optoelectronic application, due to its wide bandgap and large exciton-binding energy, has made it a very promising material for researchers across the globe. However, the difficulty of obtaining good quality, reliable and reproducible p-type ZnO has restricted the ZnO-based devices to research only, and commercial availability remains limited. The present research is focused on achieving good quality, stable p-type ZnO films by the implantation technique, which would facilitate selective area doping of the material.

PLD-deposited materials are important in depositing high-quality thin films at lower temperatures, with exact stoichiometry as the target material. Moreover, by varying the various PLD parameters, the various properties of the thin films can be changed. As the basic aim of this research was to obtain good quality p-type ZnO films required to fabricate optoelectronic devices, the PLD parameters of deposition temperature and oxygen pressure have been optimised so as to achieve very low carrier concentration, a reasonable value of Hall mobility and very high optical luminescence. Improvement in the properties of the optimised ZnO film was furthered by hydrogen implantation. A two-order increase in the optical quality of the films was observed as a result of this process.

In order to achieve p-type ZnO films, CII was carried out on the PLD-deposited ZnO films followed by annealing. Lithium and phosphorus ions were implanted for the purpose. However, no type conversion could be observed in any of the samples. While the Li-implanted samples did not demonstrate an increase in the acceptor concentration, the same was observed in the phosphorus-implanted sample. Because the conventional implantation technique did not result in the required p-type any films, we tried implanting ions using PIII technique. To do so, the ZnO films were deposited using the RF sputtering method. Nitrogen and phosphorus ions were implanted using this method, which was followed by subsequent rapid

S. Nagar and S. Chakrabarti, *Optimisation of ZnO Thin Films*,
DOI 10.1007/978-981-10-0809-2_5

thermal annealing to remove the defects. A dominant acceptor peak in the PL measurement was observed in these samples, proving the type conversion of the ZnO films. The obtained p-type ZnO films were studied over a period of 9 months. The phosphorus-implanted samples started to degrade over a period of 8 months, but the nitrogen-implanted samples continued to demonstrate acceptor peaks even after 9 months, showing that the obtained sample was quite stable.

Once the stable p-type ZnO was achieved, various heterojunction and homojunction optoelectronic devices were fabricated. A p-Si/p-ZnO heterojunction device revealed detector characteristics, with the spectral response resulting in peaks at various positions. p-ZnO/n-ZnO homojunction LED was fabricated. The p-type films were obtained by both nitrogen- and phosphorus-implanted samples, and the n-type films were deposited using the optimised PLD parameters. RTEL was observed in both these devices with dominant deep-level peaks and very little peak in the NBE region.

During the course of this research, very encouraging results have been reported towards achieving p-type ZnO films which has helped in fabricating successful homojunction LEDs. We have also reported a way to increase the luminescence properties of the ZnO thin films. However, this research work is only a small part of the wide spectrum towards making ZnO as a potential optoelectronics material. Hence, further studies need to be performed in the future to visualise the dream. Based on the experiments performed, the future work in this field can be broadly divided into material and device perspectives.

From the material perspective, the various implantation parameters of the CII can be optimised so as to obtain p-type ZnO using lithium and phosphorus ions. Other group I and group V elements can also be studied using conventional ion implantation. Moreover, PIII can be used over the PLD-deposited samples to obtain p-type ZnO films and study its reliability over a period of time. Further, co-doping methods can also use boron and nitrogen or phosphorus as group III and group V elements so as to obtain p-type films.

From the device perspective, the top n-type layer should be made thinner so as to reduce light absorption. Further, the device fabrication parameters must be optimised in order to obtain RTEL in the NBE region of ZnO. Different materials like ZnMgO can be used to further improve device characteristics, increasing efficiency and lowering dark current. ZnMgO can also be used along with ZnO films to form ZnO/ZnMgO-based quantum well structures. The heterostructures based on these quantum wells can be used to fabricate optoelectronic devices such as LEDs, lasers and some electronic devices such as thin film transistors and high-electron-mobility transistors.

ZnO, as a material, has a lot of potential in the field of emerging field of UV emitters and detectors. The realisation of stable p-type ZnO and ZnO-based devices encourages further research towards achieving commercial success.

References

1. Ozgur, U., Alivov, Y.I., Liu, C., Teke, A., Reshchikov, M., Dogan, S., Avrutin, V., Cho, S.J., Morkoc, H.: A comprehensive review of ZnO materials and devices. J. Appl. Phys. **98**, 041301–041301-103 (2005)
2. Look, D.: Recent advances in ZnO materials and devices. Mater. Sci. Eng. B **80**, 383–387 (2001)
3. Phillips, J., Bowen, W., Cagin, E., Wang, W.: Comprehensive Semiconductor Science and Technology, vol. 6.03. Elsevier Science (2011)
4. Wellenius, P.: Nitrogen Doping and Ion Beam Processing of Zinc Oxide Thin Film. North Carolina State University (2006)
5. Bian, J., Li, X., Zhang, C., Chen, L., Yao, Q.: Synthesis and characterization of two-layer-structured ZnO pn homojunctions by ultrasonic spray pyrolysis. Appl. Phys. Lett. **84**, 3783–3785 (2004)
6. Yan, Y., Zhang, S., Pantelides, S.: Control of doping by impurity chemical potentials: predictions for p-type ZnO. Phys. Rev. Lett. **86**, 5723–5726 (2001)
7. Van De Walle, C.G.: Strategies for controlling the conductivity of wide band gap semiconductors. Phys. Status Solidi B. **229**:221–228 (2002)
8. Gandhi, S.K.: VLSI Fabrication Principles, Silicon and Gallium Arsenide, Second edn. Wiley, New York (2008)
9. Lee, W., Hwang, D.K., Jeong, M.C., Lee, M., Oh, M.S., Choi, W.K., Myoung, J.M.: Fabrication and properties of As-doped ZnO films grown on GaAs (0 0 1) substrates by radio frequency (rf) magnetron sputtering. Appl. Surf. Sci. **221**, 32–37 (2004)
10. Lin, C.C., Chen, S.Y., Cheng, S.Y.: Physical characteristics and photoluminescence properties of phosphorous-implanted ZnO thin films. Appl. Surf. Sci. **238**, 405–409 (2004)
11. Walukiewicz, W.: Defect formation and diffusion in heavily doped semiconductors. Phys. Rev. B **50**, 5221 (1994)
12. Van de Walle, C.G., Laks, D., Neumark, G., Pantelides, S.: First-principles calculations of solubilities and doping limits: Li, Na, and N in ZnSe. Phys. Rev. B **47**, 9425 (1993)
13. Park, C., Zhang, S., Wei, S.-H.: Origin of p-type doping difficulty in ZnO: the impurity perspective. Phys. Rev. B **66**, 073202 (2002)
14. Look, D.C., Jones, R., Sizelove, J., Garces, N.Y., Giles, N.C., Halliburton, L.E.: The path to ZnO devices: donor and acceptor dynamics. Phys. Status Solidi A. **195**:171–177 (2003)
15. Zeng, Y., Ye, Z., Xu, W., Li, D., Lu, J., Zhu, L., Zhao, B.: Dopant source choice for formation of p-type ZnO: Li acceptor. Appl. Phys. Lett. **88**:062107–062107-3 (2006)
16. Zeng, Y.-J., Ye, Z.-Z., Xu, W.-Z., Chen, L.-L., Li, D.-Y., Zhu, L.-P., Zhao, B.-H., Hu, Y.-L.: Realization of p-type ZnO films via monodoping of Li acceptor. J. Cryst. Growth **283**, 180–184 (2005)
17. Lu, J., Zhang, Y., Ye, Z., Zeng, Y., He, H., Zhu, L., Huang, J., Wang, L., Yuan, J., Zhao, B.: Control of p-and n-type conductivities in Li-doped ZnO thin films. Appl. Phys. Lett. **89**:112113–112113-3 (2006)

S. Nagar and S. Chakrabarti, *Optimisation of ZnO Thin Films*,
DOI 10.1007/978-981-10-0809-2

18. Yang, L., Ye, Z., Zhu, L., Zeng, Y., Lu, Y., Zhao, B.: fabrication of p-type zno thin films via DC reactive magnetron sputtering by using Na as the dopant source. J. electr. Mater. **36**, 498–501 (2007)
19. Lin, S., Lu, J., Ye, Z., He, H., Gu, X., Chen, L., Huang, J., Zhao, B.: p-type behavior in Na-doped ZnO films and ZnO homojunction light-emitting diodes. Solid State Commun. **148**, 25–28 (2008)
20. Lin, S., He, H., Lu, Y., Ye, Z.: Mechanism of Na-doped p-type ZnO films: suppressing Na interstitials by codoping with H and Na of appropriate concentrations. J. Appl. Phys. **106**:093508–093508-5 (2009)
21. Gupta, M.K., Sinha, N., Singh, B., Kumar, B.: Synthesis of K-doped p-type ZnO nanorods along (100) for ferroelectric and dielectric applications. Mater. Lett. **64**, 1825–1828 (2010)
22. Jun, W., Yintang, Y.: Deposition of K-doped p type ZnO thin films on (0001) Al_2O_3 substrates. Mater. Lett. **62**, 1899–1901 (2008)
23. Li, X., Keyes, B., Asher, S., Zhang, S., Wei, S.-H., Coutts, T.J., Limpijumnong, S., Van de Walle, C.G.: Hydrogen passivation effect in nitrogen-doped ZnO thin films. Appl. Phys. Lett. **86**:122107–122107-3 (2005)
24. Look, D., Reynolds, D., Litton, C., Jones, R., Eason, D., Cantwell, G.: Characterization of homoepitaxial p-type ZnO grown by molecular beam epitaxy. Appl. Phys. Lett. **81**, 1830–1832 (2002)
25. Ashrafi, A., Suemune, I., Kumano, H., Tanaka, S.: Nitrogen-doped p-type ZnO layers prepared with H_2O vapor-assisted metalorganic molecular-beam epitaxy. Jpn. J. Appl. Phys. **41**, L1281–L1284 (2002)
26. Ye, Z.Z., Lu, J.G., Chen, H.H., Zhang, Y.Z., Wang, L., Zhao, B.H., Huang, J.Y.: Preparation and characteristics of p-type ZnO films by DC reactive magnetron sputtering. J. Crys. Growth **253**, 258–264 (2003)
27. Wang, C., Ji, Z., Liu, K., Xiang, Y., Ye, Z.: p-Type ZnO thin films prepared by oxidation of Zn_3N_2 thin films deposited by DC magnetron sputtering. J. Cryst. Growth **259**, 279–281 (2003)
28. Minegishi, K., Koiwai, Y., Kikuchi, Y., Yano, K., Kasuga, M., Shimizu, A.: Growth of p-type zinc oxide films by chemical vapor deposition. Jpn. J. Appl. Phys. **36**, L1453–L1455 (1997)
29. Li, X., Yan, Y., Gessert, T., DeHart, C., Perkins, C., Young, D., Coutts, T.: p-type ZnO thin films formed by CVD reaction of diethylzinc and NO gas. Electrochem. Solid-state Lett. **6**, C56–C58 (2003)
30. Rommeluere, J., Svob, L., Jomard, F., Mimila-Arroyo, J., Lusson, A., Sallet, V., Marfaing, Y.: Electrical activity of nitrogen acceptors in ZnO films grown by metalorganic vapor phase epitaxy. Appl. Phys. Lett. **83**, 287–289 (2003)
31. Suemune, I., Ashrafi, A., Ebihara, M., Kurimoto, M., Kumano, H., Seong, T.Y., Kim, B.J., Ok, Y.W.: Epitaxial ZnO growth and p-type doping with MOMBE. Phys. Status Solidi B. **241**:640–647 (2004)
32. Xu, W., Ye, Z., Zhou, T., Zhao, B., Zhu, L., Huang, J.: Low-pressure MOCVD growth of p-type ZnO thin films by using NO as the dopant source. J. Cryst. Growth **265**, 133–136 (2004)
33. Murphy, T., Chen, D., Cagin, E., Phillips, J.: Electronic properties of ZnO epilayers grown on c-plane sapphire by plasma-assisted molecular beam epitaxy. J. Vac. Sci. Technol. B Microelectr. Nanometer Struct. **23**, 1277–1280 (2005)
34. Dadgar, A., Oleynik, N., Bläsing, J., Deiter, S., Forster, D., Bertram, F., Diez, A., Seip, M., Greiling, A., Christen, J.: Heteroepitaxy and nitrogen doping of high-quality ZnO. J. Cryst. Growth **272**, 800–804 (2004)
35. Hwang, D.K., Kim, H.S., Lim, J.H., Oh, J.Y., Yang, J.H., Park, S.J., Kim, K.K., Look, D., Park, Y.: Study of the photoluminescence of phosphorus-doped p-type ZnO thin films grown by radio-frequency magnetron sputtering. Appl. Phys. Lett. **86**:151917–151917-3 (2005)

36. Vaithianathan, V., Lee, B.T., Kim, S.S.: Pulsed-laser-deposited p-type ZnO films with phosphorus doping. J. Appl. Phys. **98**:043519–043519-4 (2005)
37. Bang, K.-H., Hwang, D.-K., Park, M.-C., Ko, Y.-D., Yun, I., Myoung, J.-M.: Formation of p-type ZnO film on InP substrate by phosphor doping. Appl. Surf. Sci. **210**, 177–182 (2003)
38. Kang, S.-H., Hwang, D.-K., Park, S.-J.: Low-resistance and highly transparent Ni/indium-tin oxide ohmic contacts to phosphorous-doped p-type ZnO. Appl. Phys. Lett. **86**:211902–211902-3 (2005)
39. Lee, J.-M., Kim, K., Tampo, H., Yamada, A., Niki, S.: Ohmic contact to phosphorous-doped ZnO using Pt/Ni/Au for p-n homojunction diode. J. Electrochem. Soc. **153**, G1047–G1050 (2006)
40. Lee, W.-J., Kang, J., Chang, K.: Defect properties and p-type doping efficiency in phosphorus-doped ZnO. Phys. Rev. B **73**, 024117 (2006)
41. Shih, Y., Chien, J., Chen, M., Yang, J., Shiojiri, M.: P-type ZnO: P films fabricated by atomic layer deposition and thermal processing. J. Electrochem. Soc. **158**, H516–H520 (2011)
42. Kim, K.K., Kim, H.S., Hwang, D.K., Lim, J.H., Park, S.J.: Realization of p-type ZnO thin films via phosphorus doping and thermal activation of the dopant. Appl. Phys. Lett. **83**, 63–65 (2003)
43. Ryu, Y., Lee, T., Leem, J., White, H.: Fabrication of homostructural ZnO p–n junctions and ohmic contacts to arsenic-doped p-type ZnO. Appl. Phys. Lett. **83**, 4032–4034 (2003)
44. Vaithianathan, V., Lee, B.T., Kim, S.S.: Preparation of As-doped p-type ZnO films using a Zn_3As_2/ZnO target with pulsed laser deposition. Appl. Phys. Lett. **86**, 62101 (2005)
45. Kang, H.S., Kim, G.H., Kim, D.L., Chang, H.W., Ahn, B.D., Lee, S.Y.: Investigation on the p-type formation mechanism of arsenic doped p-type ZnO thin film. Appl. Phys. Lett. **89**:181103–181103-3 (2006)
46. Limpijumnong, S., Zhang, S., Wei, S.H., Park, C.: Doping by large-size-mismatched impurities: the microscopic origin of arsenic-or antimony-doped p-type zinc oxide. Phys Rev Lett **92**, 155504 (2004)
47. Sun, J., Zhao, J., Liang, H., Bian, J., Hu, L., Zhang, H., Liang, X., Liu, W., Du, G.: Realization of ultraviolet electroluminescence from ZnO homojunction with n-ZnO/p-ZnO: As/GaAs structure. Appl. Phys. Lett. **90**:121128–121128-3 (2007)
48. Liu, W., Bian, J., Hu, L., Liang, H., Zang, H., Sun, J., Zhao, Z., Liu, A., Du, G.: Electroluminescence from a ZnO homojunction device grown by pulsed laser deposition. Solid state Commun. **142**, 655–658 (2007)
49. Look, D., Renlund, G., Burgener, R., Sizelove, J.: As-doped p-type ZnO produced by an evaporation/sputtering process. Appl. Phys. Lett. **85**, 5269–5271 (2004)
50. Nagar, S., Chakrabarti, S.: Evidence of p-doping in ZnO films deposited on GaAs. Thin Solid Films **518**, 4542–4545 (2010)
51. Xiu, F., Yang, Z., Mandalapu, L., Zhao, D., Liu, J., Beyermann, W.: High-mobility Sb-doped< equation> p</equation>-type ZnO by molecular-beam epitaxy. Appl. Phys. Lett. **87**:152101–152101-3 (2005)
52. Pan, X., Ye, Z., Li, J., Gu, X., Zeng, Y., He, H., Zhu, L., Che, Y.: Fabrication of Sb-doped p-type ZnO thin films by pulsed laser deposition. Appl. Surf. Sci. **253**, 5067–5069 (2007)
53. Aoki, T., Shimizu, Y., Miyake, A., Nakamura, A., Nakanishi, Y., Hatanaka, Y.: p-Type ZnO layer formation by excimer laser doping. Phys. Status Solidi B. **229**:911–914 (2002)
54. Lopatiuk-Tirpak, O., Chernyak, L., Xiu, F., Liu, J., Jang, S., Ren, F., Pearton, S., Gartsman, K., Feldman, Y., Osinsky, A.: Studies of minority carrier diffusion length increase in p-type ZnO: Sb. J. Appl. Phys. **100**:086101–086101-3 (2006)
55. Lopatiuk-Tirpak, O., Schoenfeld, W., Chernyak, L., Xiu, F., Liu, J., Jang, S., Ren, F., Pearton, S., Osinsky, A., Chow, P.: Carrier concentration dependence of acceptor activation energy in p-type ZnO. Appl. Phys. Lett. **88**:202110–202110-3 (2006)

56. Ohara, K., Seino, T., Nakamura, A., Aoki, T., Kominami, H., Nakanishi, Y., Hatanaka, Y.: Sb excimer-laser doping in ZnO films prepared by oxidation of sulfide on Si. Appl. Surf. Sci. **244**, 369–372 (2005)
57. Wang, P., Chen, N., Yin, Z., Dai, R., Bai, Y.: p-type $Zn_{1-x}Mg_xO$ films with Sb doping by radio-frequency magnetron sputtering. Appl. Phys. Lett. **89**:202102–202102-3 (2006)
58. Przeździecka, E., Kamińska, E., Pasternak, I., Piotrowska, A., Kossut, J.: Photoluminescence study of p-type ZnO: Sb prepared by thermal oxidation of the Zn-Sb starting material. Phys. Rev. B **76**, 193303 (2007)
59. Guo, W., Allenic, A., Chen, Y., Pan, X., Che, Y., Hu, Z., Liu, B.: Microstructure and properties of epitaxial antimony-doped p-type ZnO films fabricated by pulsed laser deposition. Appl. Phys. Lett. **90**:242108–242108-3 (2007)
60. Chen, L., Lu, J., Ye, Z., Lin, Y., Zhao, B., Ye, Y., Li, J., Zhu, L.: p-type behavior in In–N codoped ZnO thin films. Appl. Phys. Lett. **87**:252106–252106-3 (2005)
61. Bin, W., Yue, Z., Jiahua, M., Wenbin, S.: Ag–N dual-accept doping for the fabrication of p-type ZnO. Appl. Phys. A **94**, 715–718 (2009)
62. Nakahara, K., Takasu, H., Fons, P., Yamada, A., Iwata, K., Matsubara, K., Hunger, R., Niki, S.: Interactions between gallium and nitrogen dopants in ZnO films grown by radical-source molecular-beam epitaxy. Appl. Phys. Lett. **79**, 4139–4141 (2001)
63. Ye, J., Gu, S., Li, F., Zhu, S., Zhang, R., Shi, Y., Zheng, Y., Sun, X., Lo, G., Kwong, D.: Correlation between carrier recombination and p-type doping in P monodoped and In–P codoped ZnO epilayers. Appl. Phys. Lett. **90**:152108–152108-3 (2007)
64. Kumar, M., Kim, S.-K., Choi, S.-Y.: Formation of Al–N co-doped p-ZnO/n-Si (100) heterojunction structure by RF co-sputtering technique. Appl. Surf. Sci. **256**, 1329–1332 (2009)
65. Kumar, M., Kim, T.-H., Kim, S.-S., Lee, B.-T.: Growth of epitaxial p-type ZnO thin films by codoping of Ga and N. Appl. Phys. Lett. **89**:112103–112103-3 (2006)
66. Yuan, G., Ye, Z., Zhu, L., Qian, Q., Zhao, B., Fan, R., Perkins, C.L., Zhang, S.: Control of conduction type in Al-and N-codoped ZnO thin films. Appl. Phys. Lett. **86**:202106–202106-3 (2005)
67. Lu, J., Ye, Z., Zhuge, F., Zeng, Y., Zhao, B., Zhu, L.: p-type conduction in N-Al co-doped ZnO thin films. Appl. Phys. Lett. **85**, 3134–3135 (2004)
68. Chen, Z., Kawasuso, A., Xu, Y., Naramoto, H., Yuan, X., Sekiguchi, T., Suzuki, R., Ohdaira, T.: Production and recovery of defects in phosphorus-implanted ZnO. J. Appl. Phys. **97**:013528–013528-6 (2005)
69. Lin, C.C., Chen, S.Y., Cheng, S.Y., Lee, H.Y.: Properties of nitrogen-implanted p-type ZnO films grown on SiN/Si by radio-frequency magnetron sputtering. Appl. Phys. Lett. **84**, 5040 (2004)
70. Braunstein, G., Muraviev, A., Saxena, H., Dhere, N., Richter, V., Kalish, R.: p type doping of zinc oxide by arsenic ion implantation. Appl. Phys. Lett. **87**:192103–192103-3 (2005)
71. Vaithianathan, V., Hishita, S., Park, J.Y., Kim, S.S.: Photoluminescence in phosphorous-implanted ZnO films. J. Appl. Phys. **102**:086107–086107-3 (2007)
72. Teke, A., Özgür, Ü., Doğan, S., Gu, X., Morkoç, H., Nemeth, B., Nause, J., Everitt, H.: Excitonic fine structure and recombination dynamics in single-crystalline ZnO. Phys. Rev. B **70**, 195207 (2004)
73. Alivov, Y.I., Johnstone, D., Özgür, Ü., Avrutin, V., Fan, Q., Akarca-Biyikli, S., Morkoç, H.: Electrical and optical properties of n-ZnO/p-SiC heterojunctions. Jpn. J. Appl. Phys. **44**, 7281 (2005)
74. Ye, J., Gu, S., Zhu, S., Liu, W., Liu, S., Zhang, R., Shi, Y., Zheng, Y.: Electroluminescent and transport mechanisms of n-ZnO/p-Si heterojunctions. Appl. Phys. Lett. **88**, 182112 (2006)
75. Chen, P., Ma, X., Yang, D.: Ultraviolet electroluminescence from ZnO/p-Si heterojunctions. J. Appl. Phys. **101**:053103–053103-4 (2007)

76. Tsukazaki, A., Ohtomo, A., Onuma, T., Ohtani, M., Makino, T., Sumiya, M., Ohtani, K., Chichibu, S.F., Fuke, S., Segawa, Y.: Repeated temperature modulation epitaxy for p-type doping and light-emitting diode based on ZnO. Nat. Mater. **4**, 42–46 (2004)
77. Lim, J.H., Kang, C.K., Kim, K.K., Park, I.K., Hwang, D.K., Park, S.J.: UV electroluminescence emission from ZnO light emitting diodes grown by high temperature radiofrequency sputtering. Adv. Mater. **18**, 2720–2724 (2006)
78. Alivov, Y.I., Van Nostrand, J., Look, D.C., Chukichev, M., Ataev, B.: Observation of 430 nm electroluminescence from ZnO/GaN heterojunction light-emitting diodes. Appl. Phys. Lett. **83**, 2943–2945 (2003)
79. Alivov, Y.I., Kalinina, E., Cherenkov, A., Look, D.C., Ataev, B., Omaev, A., Chukichev, M., Bagnall, D.: Fabrication and characterization of n-ZnO/p-AlGaN heterojunction light-emitting diodes on 6H-SiC substrates. Appl. Phys. Lett. **83**, 4719 (2003)
80. Osinsky, A., Dong, J., Kauser, M., Hertog, B., Dabiran, A., Chow, P., Pearton, S., Lopatiuk, O., Chernyak, L.: MgZnO/AlGaN heterostructure light-emitting diodes. Appl. Phys. Lett. **85**, 4272 (2004)
81. Chuang, R.W., Wu, R.-X., Lai, L.-W., Lee, C.-T.: ZnO-on-GaN heterojunction light-emitting diode grown by vapor cooling condensation technique. Appl. Phys. Lett. **91**:231113–231113-3 (2007)
82. Xu, H.Y., Liu, Y., Liu, Y., Xu, C., Shao, C., Mu, R.: Ultraviolet electroluminescence from p-GaN/i-ZnO/n-ZnO heterojunction light-emitting diodes. Appl. Phys. B **80**, 871–874 (2005)
83. Kim, D.C., Han, W.S., Cho, H.K., Kong, B.H., Kim, H.S.: Multidimensional ZnO light-emitting diode structures grown by metal organic chemical vapor deposition on p-Si. Appl. Phys. Lett. **91**:231901–231901-3 (2007)
84. Nakamura, A., Ohashi, T., Yamamoto, K., Ishihara, J., Aoki, T., Temmyo, J., Gotoh, H.: Full-color electroluminescence from ZnO-based heterojunction diodes. Appl. Phys. Lett. **90**:093512–093512-3 (2007)
85. Xi, Y., Hsu, Y., Djurišić, A., Ng, A., Chan, W., Tam, H., Cheah, K.: NiO/ZnO light emitting diodes by solution-based growth. Appl. Phys. Lett. **92**, 113505 (2008)
86. Ohta, H., Kawamura, K.-I., Orita, M., Hirano, M., Sarukura, N., Hosono, H.: Current injection emission from a transparent p–n junction composed of p-$SrCu_2O_2$/n-ZnO. Appl. Phys. Lett. **77**, 475–477 (2000)
87. Tsukazaki, A., Kubota, M., Ohtomo, A., Onuma, T., Ohtani, K., Ohno, H., Chichibu, S.F., Kawasaki, M.: Blue light-emitting diode based on ZnO. arXiv preprint cond-mat/0504587 (2005)
88. Liu, W., Gu, S., Ye, J., Zhu, S., Liu, S., Zhou, X., Zhang, R., Shi, Y., Zheng, Y., Hang, Y.: Blue-yellow ZnO homostructural light-emitting diode realized by metalorganic chemical vapor deposition technique. Appl. Phys. Lett. **88**:092101–092101-3 (2006)
89. Neamen, D.A.: Semiconductor Physics and Devices. McGraw Hill, Boston (2003)
90. Ievtushenko, A., Lashkarev, G., Lazorenko, V., Karpyna, V., Sichkovskyi, V., Kosyachenko, L., Rchuk, V., Klyarchuk, O., Bosy, V., Korzhinski, F.: Ultraviolet detectors based on ZnO:N thin films with different contact structures. Acta Phys. Polonica A **114**, 1123–1129 (2008)
91. Fabricius, H., Skettrup, T., Bisgaard, P.: Ultraviolet detectors in thin sputtered ZnO films. Appl. Opt. **25**, 2764–2767 (1986)
92. Zheng, X., Li, Q.S., Zhao, J., Chen, D., Zhao, B., Yang, Y., Zhang, L.C.: Photoconductive ultraviolet detectors based on ZnO films. Appl. Surf. Sci. **253**, 2264–2267 (2006)
93. Zhang, D., Brodie, D.: Photoresponse of polycrystalline ZnO films deposited by rf bias sputtering. Thin Solid Films **261**, 334–339 (1995)
94. Pillai, P., Pillai, C., Mendiratta, R.: Space-charge polarisation in ZnO photoconductors. J. Phys. D Appl. Phys. **12**, 961 (1979)

95. Mandalapu, L., Xiu, F., Yang, Z., Liu, J.: Ultraviolet photoconductive detectors based on Ga-doped ZnO films grown by molecular-beam epitaxy. Solid-state Electr. **51**, 1014–1017 (2007)
96. Ryu, Y., Lee, T., Lubguban, J., White, H., Park, Y., Youn, C.: ZnO devices: Photodiodes and p-type field-effect transistors. Appl. Phys. Lett. **87**:153504–153504-3 (2005)
97. Suhaila, A., Salmana, S., Naouma, F., Thjeela, H.: ZnO/Porous-silicon photovoltaic UV detector. J. Electr. Devices **13**, 900–909 (2012)
98. Jeong, I.-S., Kim, J.H., Im, S.: Ultraviolet-enhanced photodiode employing n-ZnO/p-Si structure. Appl. Phys. Lett. **83**, 2946–2948 (2003)
99. Wang, K., Vygranenko, Y., Nathan, A.: Optically transparent ZnO-based n–i–p ultraviolet photodetectors. Thin Solid Films **515**, 6981–6985 (2007)
100. Wang, K., Vygranenko, Y., Nathan, A.: ZnO-based pin and nip heterostructure ultraviolet sensors: a comparative study. J. Appl. Phys. **101**:114508–114508-5 (2007)
101. Ohta, H., Kamiya, M., Kamiya, T., Hirano, M., Hosono, H.: UV-detector based on pn-heterojunction diode composed of transparent oxide semiconductors, p-NiO/n-ZnO. Thin Solid Films **445**, 317–321 (2003)
102. Tapily, K., Gu, D., Baumgart, H.: Growth Mechanism of ALD ZnO Films Investigated by Physical Characterization. ECS Transactions **33**, 355–363 (2010)
103. Paraguay, D.F., Estrada, L.W., Acosta, N.D., Andrade, E., Miki-Yoshida, M.: Growth, structure and optical characterization of high quality ZnO thin films obtained by spray pyrolysis. Thin Solid Films. **350**:192–202 (1999)
104. Meyer, B., Alves, H., Hofmann, D., Kriegseis, W., Forster, D., Bertram, F., Christen, J., Hoffmann, A., Straßburg, M., Dworzak, M.: Bound exciton and donor–acceptor pair recombinations in ZnO. Phys. Status Solidi B. **241**:231–260 (2004)
105. Wagner, M., Callsen, G., Reparaz, J., Schulze, J.H., Kirste, R., Cobet, M., Ostapenko, I., Rodt, S., Nenstiel, C., Kaiser, M.: Bound excitons in ZnO: Structural defect complexes versus shallow impurity centers. Phys. Rev. B **84**, 035313 (2011)
106. Hsieh, P.T., Chen, Y.C., Wang, C.M., Tsai, Y.Z., Hu, C.C.: Structural and photoluminescence characteristics of ZnO films by room temperature sputtering and rapid thermal annealing process. Appl. Phys. A. Mater. Sci. Process. **84**, 345–349 (2006)
107. Wang, X., Buyanova, I., Chen, W., Pan, C., Tu, C.: Effects of stoichiometry on defect formation in ZnO epilayers grown by molecular-beam epitaxy: An optically detected magnetic resonance study. J. Appl. Phys. **103**:023712–023712-4 (2008)
108. Misra, P., Sharma, T., Kukreja, L.: Temperature dependent photoluminescence processes in ZnO thin films grown on sapphire by pulsed laser deposition. Curr. Appl. Phys. **9**, 179–183 (2009)
109. Jung, S., Park, W., Cheong, H., Yi, G.C., Jang, H.M., Hong, S., Joo, T.: Time-resolved and time-integrated photoluminescence in ZnO epilayers grown on Al_2O_3 (00.1) by metalorganic vapor phase epitaxy. Appl. Phys. Lett **80**, 1924 (2002)
110. Thomas, D., Lander, J.: Hydrogen as a donor in zinc oxide. J. Chem. Phys. **25**, 1136 (1956)
111. Van de Walle, C.G.: Hydrogen as a cause of doping in zinc oxide. Physical review letters **85**, 1012 (2000)
112. Cox, S., Davis, E., Cottrell, S., King, P., Lord, J., Gil, J., Alberto, H., Vilao, R., Duarte, J. P., de Campos, N.A.: Experimental confirmation of the predicted shallow donor hydrogen state in zinc oxide. Phys. Rev. Lett. **86**, 2601 (2001)
113. Lee, J.-K., Nastasi, M., Hamby, D., Lucca, D.: Optical observation of donor-bound excitons in hydrogen-implanted ZnO. Appl. Phys. Lett. **86**:171102–171102-3 (2005)
114. Fang, Z.-Q., Claflin, B., Look, D.C., Farlow, G.C.: Electron irradiation induced deep centers in hydrothermally grown ZnO. J. Appl. Phys. **101**:086106–086106-3 (2007)
115. Simpson, J., Cordaro, J.F.: Characterization of deep levels in zinc oxide. J. Appl. Phys. **63**, 1781–1783 (1988)

116. Kuriyama, K., Ooi, M., Matsumoto, K., Kushida, K.: Thermally stimulated current studies on deep levels in hydrothermally grown single crystal ZnO bulk. Appl. Phys. Lett. **89**:242113–242113-3 (2006)
117. Fang, Z.-Q., Claflin, B., Look, D.C.: Effects of annealing in N_2 ambient on traps and persistent conduction in hydrothermally grown ZnO. J. Appl. Phys. **103**:073714–073714-5 (2008)
118. Lopatiuk, O., Chernyak, L., Osinsky, A., Xie, J.: Lithium-related states as deep electron traps in ZnO. Appl. Phys. Lett. **87**:214110–214110-3 (2005)
119. Wardle, M., Goss, J., Briddon, P.: Theory of Li in ZnO: A limitation for Li-based p-type doping. Phys. Rev. B **71**, 155205 (2005)
120. Zhao, D., Xu, S., Xie, M., Tong, S., Yang, H.: Stress and its effect on optical properties of GaN epilayers grown on Si (111), 6H-SiC (0001), and c-plane sapphire. Appl. Phys. Lett. **83**, 677–679 (2003)
121. Sun, J., Lu, Y., Liu, Y., Shen, D., Zhang, Z., Li, B., Zhang, J., Yao, B., Zhao, D., Fan, X.: The activation energy of the nitrogen acceptor in p-type ZnO film grown by plasma-assisted molecular beam epitaxy. Solid State Commun. **140**, 345–348 (2006)
122. Barnes, T.M., Olson, K., Wolden, C.A.: On the formation and stability of p-type conductivity in nitrogen-doped zinc oxide. Appl. Phys. Lett. **86**:112112–112112-3 (2005)
123. Lu, J., Zhang, Y., Ye, Z., Zhu, L., Wang, L., Zhao, B., Liang, Q.: Low-resistivity, stable p-type ZnO thin films realized using a Li–N dual-acceptor doping method. Appl. Phys. Lett. **88**:222114–222114-3 (2006)
124. Allenic, A., Guo, W., Chen, Y., Katz, M.B., Zhao, G., Che, Y., Hu, Z., Liu, B., Zhang, S. B., Pan, X.: Amphoteric phosphorus doping for stable p-type ZnO. Adv. Mater. **19**, 3333–3337 (2007)
125. Ryu, Y., Kim, W., White, H.: Fabrication of homostructural ZnO p–n junctions. J. Cryst. Growth **219**, 419–422 (2000)
126. Xiong, G., Wilkinson, J., Mischuck, B., Tuzemen, S., Ucer, K., Williams, R.: Control of p-and n-type conductivity in sputter deposition of undoped ZnO. Appl. Phys. Lett. **80**, 1195–1197 (2002)
127. Liang, H., Feng, Q., Sun, J., Zhao, J., Bian, J., Hu, L., Zhang, H., Luo, Y., Du, G.: Room temperature electroluminescence from the ZnO homojunction grown on an n+-Si substrate by metal–organic chemical vapor deposition. Semicond. Sci. Technol. **23**, 025014 (2008)
128. Du, G., Liu, W., Bian, J., Hu, L., Liang, H., Wang, X., Liu, A., Yang, T.: Room temperature defect related electroluminescence from ZnO homojunctions grown by ultrasonic spray pyrolysis. Appl. Phys. Lett. **89**:052113–052113-3 (2006)

MIX
Papier aus verantwortungsvollen Quellen
Paper from responsible sources
FSC® C105338

If you have any concerns about our products,
you can contact us on
ProductSafety@springernature.com

In case Publisher is established outside the EU,
the EU authorized representative is:
Springer Nature Customer Service Center GmbH
Europaplatz 3, 69115 Heidelberg, Germany

Printed by Libri Plureos GmbH
in Hamburg, Germany